Pharmaceutical Metrics

To Lois, Rachael, Jeffrey and Aaron

Pharmaceutical Metrics

Measuring and Improving R & D Performance

David S. Zuckerman

GOWER

Published by
Gower Publishing Limited
Gower House
Croft Road
Aldershot
Hampshire GU11 3HR
England

Gower Publishing Company
Suite 420
101 Cherry Street
Burlington, VT 05401–4405
USA

David S. Zuckerman has asserted his right under the Copyright, Designs and Patents Act 1988 to be identified as the author of this work.

British Library Cataloguing in Publication Data
Zuckerman, David S.
    Pharmaceutical metrics: measuring and improving R & D
    performance
    1. Drug development – management 2. Drugs – Research –
    Management 3. Drugs – Analysis – Methodology
    4. Pharmaceutical industry – Management 5. Pharmaceutical
    industry – Quality control
    I. Title
    615.1'9'00724

    ISBN 0 566 08676 X

Library of Congress Cataloging-in-Publication Data
Zuckerman, David S.
    Pharmaceutical metrics: measuring and improving R&D performance/
    David S. Zuckerman
      p.cm.
    ISBN 0-566-08676-X
     1. Pharmaceutical industry--Research--Statistical methods. 2.
Pharmaceutical industry--Marketing--Research--Statistical methods. 3.
Drugs--Marketing--Research--Statistical methods. I. Title.

HD9665.5Z83 2006
615.1'068'5--dc22

2005053608

Typeset in Bembo by IML Typographers, Birkenhead and printed
in Great Britain by MPG Books Ltd, Bodmin, Cornwall

Contents

List of Figures

LIST OF
FIGURES

List of Abbreviations

ACWP	actual cost of work performed
AE	adverse event
b	billion
BAC	budget at completion
BCWP	budgeted cost of work performed
BCWS	budgeted cost of work scheduled
BSC	balanced scorecard
CEO	chief executive officer
CMC	chemistry, manufacturing and controls
COO	chief operating officer
CPI	cost performance index
CRA	contract research associate
CRF	case report form
CRO	contract research organization
CT	cycle time
CV	cost variance
DBC	database closed
DBL	database locked
Dvmt	development
E	efficiency
EAC	estimate at completion
EDC	electronic data capture
EV	earned value
EVA	earned value analysis
FAA	Federal Aviation Administration (USA)
FDA	Food & Drug Administration (USA)
FMEA	failure modes and effects analysis
FPFV	first patient first visit
FR	final report
FRD	final report draft
G	green
HPLC	high performance liquid chromatography
HR	human resources
IM	information management
LPLV	last patient last visit
m	million

MA	moving average
MBNQA	Malcolm Baldrige National Quality Award (USA)
n/a	not applicable
NCE	new chemical entity
NDA	new drug application
NOAC	next operation as customer
obs	observation
OTC	over the counter
PDCA	Plan-Do-Check-Act
Q	quality
QA	quality assurance
R	red
R & D	Research and Development
RA	regulatory affairs
ROA	return on assets
ROI	return on investment
SA	statistical analysis
SAC	schedule at completion
SAE	serious adverse event
SPI	schedule performance index
SV	schedule variance
T	timeliness
TA	therapeutic area
TLG	tables, listings and graphs
VAC	variance at completion
Y	yellow

Preface

I have been fortunate to work with clients in the pharmaceutical industry for over ten years. The industry is populated with bright, inquisitive, highly educated people who never shrink from a challenge. This has made my work highly enjoyable and stimulating. Whether I've been working with clients on process improvements, teambuilding, change management or other improvement projects, I've found both companies and individuals receptive to new ideas and approaches.

However, pharmaceutical R & D has always been an operation that has operated on instinct and intuition. As the industry has matured and competition has stiffened, the need for a more systematic, rigorous approach to R & D has become evident. I've considered it both a challenge and a privilege to lead the charge toward measuring and improving the way the industry accomplishes drug development. My goal has been to preserve the intuition while systematizing the processes that surround and support it.

I hope that this book will serve that goal by providing insights into how to measure and improve the overall R & D process. Like the science we practice, our organizations can't improve unless measurement takes place.

I have provided four case studies – example companies – that I use to illustrate the concepts developed in this book. Each of these companies is entirely fictional, but drawn from my many years of experience in the field. To my knowledge, there is no actual company that resembles any of these case studies, although readers may find that parts of each case study parallel their own experience.

Acknowledgements

Writing a book turned out to be much more complicated than I ever imagined, and it took the efforts of many people to help me complete it. I'd like to thank Jonathan Norman, Publisher for Gower, for first approaching me about writing this book and cheering me on throughout the process. He was ably assisted by Fiona Martin in developing the manuscript. Nikki Dines and Sarah Norman provided tremendous help during the editing process. Claire Percy, Sue White and Gemma Court made the publication marketing process seem positively easy. My thanks to all of them.

Meanwhile, I must give great credit to Drs Gen Li and Charles Piper for their insights and help with various technical aspects of the book. Gen read the entire manuscript to ensure that everything made sense and was consistent. Chuck added several examples that brought out the key points in the book. I am very grateful for their assistance. Any remaining mistakes in the book are entirely mine.

Finally, my many clients allowed me to refine the examples, test new concepts and develop new methods. Providing balanced scorecard classes for the Institute of International Research, Pharmaceutical Training International Division (PTI) let me try out the various case studies and hone the descriptions. I greatly appreciate the opportunities they provided.

Many thanks to all who helped so much in this endeavor.

CHAPTER 1

The Case for Metrics

The pharmaceutical business is incredibly complex. It's hard to keep track of all the different things that have to be done: find new molecular entities that will treat disease, prove that these compounds are safe and effective, bring them to market, manufacture them safely and consistently, and work with regulatory officials in dozens of different countries. And, in the meantime, make a profit while getting all this done. And that's just what the senior executives have to worry about! The chemists, accountants and technicians that do the day-to-day work have so many more details to consider. It can be completely overwhelming.

But that's exactly the point. With so many different considerations to worry about every day, it's hard to know what's important and what's not; where we should spend our time and energy and what we should avoid. The problem is so complex, maybe we should just continue about our business and let someone else worry about the big picture.

Then again, if we could focus just a little bit more, maybe we could get more work done faster. And doing the right things faster is critical. The current estimate of the cost to bring a drug to market has been estimated by some to be over $800 m and 10–12 years.[1] How much of that time and money is wasted? How much is spent on things that don't really need to be done, could be done faster and more efficiently, or simply aren't as important to do right now? Here's a simple example: Most pharmas currently use many contract research organizations (CROs) to help with their drug development. Each new CRO that a pharma brings in has to learn that pharma's way of doing business, clinical trial software, and so on. Meanwhile, the pharma must learn about the CRO's way of doing things. There's a 'learning curve' associated with doing new things with new suppliers (see Figure 1.1). The second time pharma and CRO work together things go more quickly (everything from negotiating the contract to sending documents to the right reviewers). The third time, things go even faster. Each time we repeat the same work with the same CRO, things get faster, more efficient and less expensive. In fact, if we work together long enough, some tasks such as contract negotiations can perhaps be completely eliminated. So there's a real time and cost saving in working with the same CRO multiple times.

---

1. 'The Price of Innovation: New Estimates of Drug Development Costs', *Journal of Health Economics*, J.A. DiMasi, R.W. Hansen, H.G. Grabowski, Mar 2003, 22(2): 151–85.

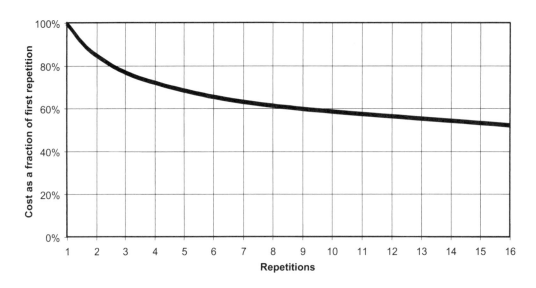

Figure 1.1 Performing the same task repeatedly (e.g., working repeatedly with a single CRO) yields learning curve benefits. In this example, each doubling on repetitions reduces the cost by 15% (85% learning curve)

However, if we don't think about all of the time and cost associated with each new CRO relationship, we will – and do – waste incredible amounts of time and money. It's not unusual for a pharma to work with a different CRO on every project. So the learning curve benefits shown in Figure 1.1 are never ever achieved!

That's what metrics are all about. If we could somehow measure what we do and compare that to what we really need to do, maybe we can increase the amount of time we spend on productive work, and eliminate some of the wasted work. Maybe we can figure out ways to eliminate tasks that simply don't need to be done or compress tasks that aren't really adding value. As regards a pharma and its CROs, we might find that we make savings on each project by making multiple CROs compete, but the overall cost of switching CROs from one project to the next far exceeds that saving. Maybe; maybe not. But until we measure, there's no way to know for sure.

Now you may be saying to yourself: 'This is just a way to work harder and faster, right?' Well, I'll agree with you that measuring and improving tasks will help you work *faster*, but I'd prefer *smarter* to *harder*. Let me give you an example:

Some years ago Price Waterhouse did an analysis of how effectively people work.[2] They divided work up into two categories, which I'll call 'effective' and 'ineffective'. Effective work is work that materially contributes to successfully delivering something, such as analyzing a sample or administering a drug to a patient. Ineffective work, on the other hand, doesn't materially contribute. Examples of ineffective work include:

---

2. Price Waterhouse popularized the concept of effective work in the early 1990s and often cited white-collar values for ineffective work as high as 95 percent. This information was provided primarily in their marketing and consulting materials. No published documentation of effective/ineffective values is available, but experience tends to bear out these high ineffective values.

- waiting (for something to happen or someone else to arrive or perform their task)
- transporting something (from one place to another)
- fixing mistakes (rework)
- inspecting (to make sure someone else didn't make a mistake).

The theory is that if you're doing any of these ineffective things, you're not materially contributing to successful delivery, so from the product standpoint you're being ineffective. Look at the example in Figure 1.2. You go to a restaurant and order some toast. First you wait for the server to come to your table. Then you give the server the order. The server then gives the order to the cook, who walks over to the counter, picks up two pieces of toast, walks over to the toaster and puts them in. The toaster does its job and pops up when it's done. The cook, who is now busy filling another order, eventually returns to the toaster and inspects the toast: it's not burnt, so it goes on a plate and is set out for the server. The server has also been off helping other customers, but eventually returns, picks up your toast and brings it to your table. You inspect the toast and realize that it's cold (since it's been sitting around waiting for the cook and the server), so you send it back and the whole process is repeated. Everyone is more vigilant the second time and the toast appears hot and delicious.

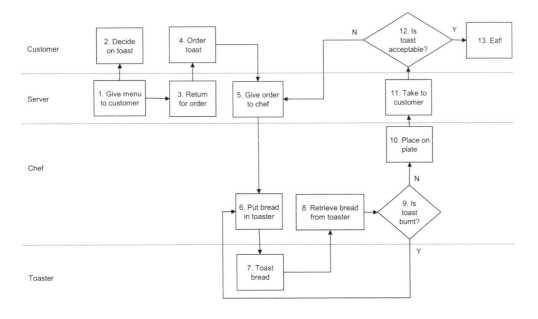

Figure 1.2   A process for ordering toast

Spend a minute with Figure 1.2 and count the number of 'effective' tasks. If you found only two, you're right. The only truly effective tasks are the actual toasting and eating (Tasks 7 and 13). OK, we could also count the task of telling the server your order, since none of us are clairvoyant (I suspect that if the server could really read your mind, he or she wouldn't remain a server for long!). I'll even let you count putting the toast in the toaster and maybe even putting it on a plate. Even so, the number of ineffective tasks is staggering: at least 54 percent of the toast tasks (7 out of 13) are ineffective (see Figure 1.3), and are included simply because of the way the restaurant is set up. If the server could make the toast and it popped up right at your table, all of the other tasks would disappear and you'd have a much better chance of getting hot toast the first time!

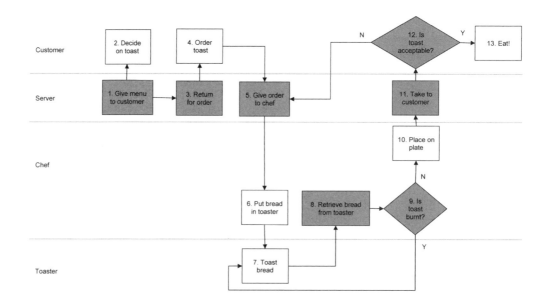

Figure 1.3   Most of the toast ordering process is ineffective (ineffective tasks are shaded)

In fact, when Price Waterhouse examined a wide variety of office and research processes, they found that a staggering 80 to 95 percent of the work we perform falls into the 'ineffective' category! Don't believe it? Try it on your own work some day. Simply keep track of the tasks you do (or just keep track of the time you spend doing 'effective' and 'ineffective' tasks during a given period of time) and you'll be amazed.

Before you get too depressed about how ineffective we all are, consider this. It's not possible – or even desirable – to remove all ineffective work; otherwise we'd just be robots going about our individual tasks. Human enterprise requires that we interact, organize, socialize. True invention and breakthroughs come from these complex interactions. So while we're incredibly inefficient at delivering things, those inefficiencies yield other, extremely valuable, results. But do we have to be 95 percent ineffective? If we could even cut our ineffective work from 95 to 90 percent, we would have doubled our effective work from 5 to 10 percent! So there's great opportunity in focusing on effective work and removing even a small amount of ineffective work, because the gains in effective work can be staggering.

But who's to say what effective work is? If you're delivering toast, it's relatively simple. But if you're working in the research division of a pharmaceutical company it's a lot more complicated. Say you're a clinical pathologist. You receive samples from a research biologist colleague in another department with an analysis request. You don't have the right equipment for the analysis, but your colleague knows and trusts you and wasn't sure who to give it to. You forward it to a biologist who performs the analysis and sends it back to your colleague. You get a frustrated call saying the analysis wasn't done correctly and you have to mediate a heated discussion before the matter is ironed out and the analysis is performed to the requester's needs. In fact, Figure 1.4 shows that your entire participation was ineffective because all you did was transmit and mediate; you did no actual work (although you did mend some political fences and maintain harmony in the department,

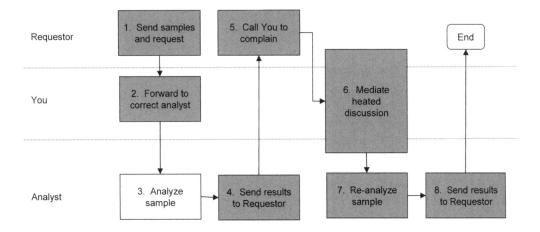

Figure 1.4 Most white-collar processes have large amounts of ineffective work (ineffective tasks are shaded)

which is no mean feat). If your colleague had known who to go to in the first place and contacted that person directly, plenty of time and energy could have been saved.

So the purpose of metrics is this:

to measure effective progress toward our goals.

We define effective work by identifying the goals we want to attain. We then can create specific metrics that allow us to measure how much of our effort is successfully, *effectively* directed toward those goals.

If you like sports analogies, you can think of metrics in terms of a game of American football. During each play we hope to move the ball down the field and keep the opposing team from doing the same. Progress is sometimes measured in tens of yards, and sometimes in inches. Sometimes we move backwards or lose control of the ball altogether. Eventually, we reach the end zone and tally points. Then it starts again. Winning in football is about measuring our progress toward the goal line and ignoring anything else that might distract us (cheerleaders, fans and maybe even coaches!). It's a simple analogy, but the point is clear. Move the ball and ignore the distractions, score some points and do it again, all the while preventing your adversary from doing the same thing.

THREE PRINCIPLES OF MEASUREMENT

As we continue our discussions, you should keep in mind three principles of good metrics systems:

1   Metrics measure progress toward our goals.
2   Metrics help us prevent failure.
3   Metrics get us out of ruts.

Principle 1: Metrics measure progress toward our goals

First, metrics measure progress toward our goals. By defining what our goals are, and then identifying measurements that will help us attain them, we can use a metrics system to help us progress. Figure 1.5 shows the progression.

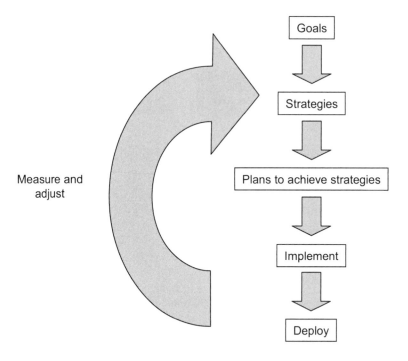

Figure 1.5   Results are accomplished when goals drive actions and metrics measure progress

Too often, we look around and see a bunch of things we can measure and then say 'Great. Now I've got a metrics system!' But the things we're measuring may or may not have anything to do with what we're trying to achieve.

The simplest analogy here is an automobile dashboard. When you're driving, you're most interested in three things: speed, direction and range (how far can I get on the remaining gas I've got in the tank). Every car has a speedometer and gas gauge, which are essential. They often also have tachometers (which haven't been useful since the days of manual transmissions). Knowing the engine temperature is nice, but only necessary in extreme circumstances. I can't remember the last time I needed to know the battery voltage. Beyond that, how many of those myriad dials and readouts do you really need? Does your car have a compass? A 'distance to empty' readout that tells you how far you can go before you run out of gas? When it comes to metrics, your car's dashboard has some (but usually not all) of the important measures, and lots of measures most of us rarely need or use.

Pharmaceutical R & D is much the same. It's easy to measure whether we've hit a forecasted milestone on time, so most R & D metrics systems measure that. It's nice information to know, but only if getting done on time is what's important. In the drug development business, getting done on time is fine, but the real goal is to *get the job done right in progressively shorter amounts of time and at progressively lower cost.* So the important measures are cycle time (how long it takes us to do the task) and quality (how good is the product we've produced). Like our cars, just because we *can* measure something doesn't mean we *should* measure it. Figure out your development goals and create measures that give your speed, direction and range relative to those goals.

So Figure 1.5 shows us that we have to establish our vision, goals and objectives before we establish our metrics. That way we're measuring progress toward our goals, not just measuring for measurement's sake. In Chapter 3 we'll talk about how to build vision, mission and objectives for your organization, and then how to tie your metrics to the resulting strategies.

Principle 2: Metrics help us prevent failure

It's not enough simply to measure progress toward our goals. We have to be able to identify and prevent failures as well. Otherwise we run the risk of making the same mistakes again and again; thereby preventing us from ultimately achieving our goal. The Titanic crew was so sure of their ability to achieve the goal of crossing the Atlantic in record time, and so convinced of their ship's invincibility, that they never really paid attention to the iceberg that sunk the ship! If you don't learn how to prevent – and recover from – failure, you can never achieve your goals.

Figure 1.6a shows the problem we face when we're working toward a goal. As we're doing our jobs, events outside of our control intervene. These events could be anything from a sudden, high-priority customer request, to unexpectedly poor clinical trial enrollment, to a failure in trying to deliver a product to a customer. We're forced to compensate for the problem in some way: take care of the high-priority request before getting back to our scheduled work, figure out how to get sites to enroll better so we can get enrollment back on track, or try and figure out what went wrong in the product delivery and how to get the right product to the customer. In any case, we're forced to make everything else wait while we become 'fire fighters' and take care of the panic (Figure 1.6b).

I'm sure you've been in this situation on many occasions. I certainly have. Sometimes the fire fighting becomes so bad that we spend all of our time just putting out one fire after another; we never manage to get back to the real work we need to do to achieve our goals.

Here's where metrics come in. Most of these 'outside events' are to some extent predictable. The customer that comes to you with a rush request usually knew in advance that he or she would need this work done; enrollment failures are so common that they are actually highly predictable; and product delivery failures often come in batches as a result of a process problem. Without measurement, every problem looks unique and unpredictable. But with measurement, we start to see the trends and understand what's causing them. That allows us to get out of the 'fire fighter' loop and into the 'corrective action' loop[3] (see Figure 1.6c). Even more important, we can use the completion of each

---

3. 'The fire fighter/corrective action loop concept is based on the idea of multi-loop learning, first suggested by Chris Argyris in 'Good Communication that Blocks Learning', *Harvard Business Review*, July/Aug 1994.

THE CASE FOR
METRICS

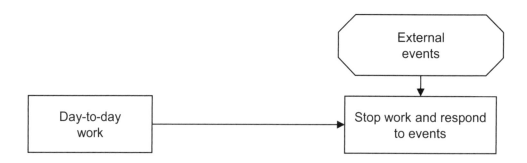

Figure 1.6a   External events often disrupt work and require immediate, remedial action

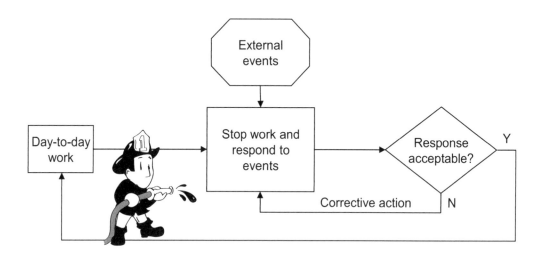

Figure 1.6b   We then become 'fire fighters' who are responding to crises more often than getting effective work accomplished

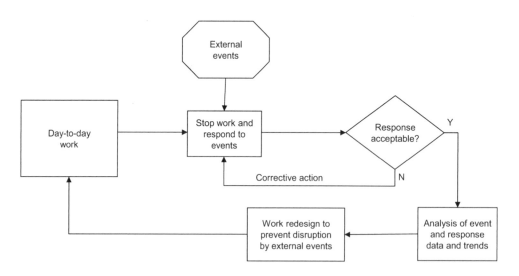

Figure 1.6c   By measuring performance and trends, we can take preventive action and begin to head off problems rather than just responding

study to measure our success and define best practices and lessons learned (no matter how painful) for use on future projects. Once we're in these larger loops, we can actually change the design of our systems to prevent or accommodate these interfering 'outside events'. We can build a request system that allows customers to alert us of impending, high-priority requests. We can use new enrollment forecasting tools to figure out how to make our protocols 'underenrollment' proof. We can use failure modes and effects analysis (FMEA) to predict delivery failures and come up with contingency and prevention plans.

Principle 3: Metrics get us out of ruts

One of the maxims I use in my business is this:

The definition of insanity is doing the same things and expecting different results.

Most organizations and businesses fall into this trap, and pharmaceutical R & D is no exception. Clinical trial enrollment is a good example. Virtually every clinical trial runs into enrollment problems. This is due to various problems, including overly restrictive inclusion/exclusion criteria and overly optimistic investigators. If you've ever been associated with a clinical trial, I'm not telling you anything new. Yet, few organizations have taken the time and effort to track the various costs, causes and cures for enrollment problems and systematically take action to improve things. Instead, R & D organizations tend to rely on doing more of the same: inspecting their protocols more carefully, doing cursory enrollment feasibility analyses and hoping that their investigators live up to their optimistic forecasts. And predictably, the results are the same: slow enrollment, late enrollment and panic. Ask any enrollment consultant where they do most of their business, and they'll tell you it's in helping clinical trial teams get enrollment back on track after things have gone awry. Very little of their business comes from helping clinical trial teams get their protocols right in the first place. However, if the organization were to carefully measure enrollment costs, causes and cures, they would probably find that an ounce of prevention is worth a pound of cure; spending some time and money up front to prevent enrollment problems would be much more effective than spending time and money later to fix them.

EXAMPLE COMPANIES

Throughout this book, I'll draw on four example companies. Each of these companies is entirely fictional, and I've developed each in order to illustrate a different aspect of the industry:

- *Inventrix* illustrates metrics for a big pharma.
- *BioStart* illustrates metrics for a micro pharma.
- *VirtuPharm* illustrates metrics for a mid-size pharma attempting to do their development virtually.
- *CROmed* illustrates metrics for a CRO.

It is likely you will find aspects of each case study company that are applicable to your situation.

Inventrix

Inventrix is a new entity formed by the 'big pharma' company Gargantua. Gargantua took various late life cycle drugs (some prescription, some over the counter, OTC), some state-of-the-art manufacturing technologies and production facilities, a handful of development labs and about 2000 people, and created *Inventrix*. The goal of the new company is to use the existing drugs as a funding source and develop new product revenue streams that can contribute to the company's bottom line.

Gargantua had been attempting to become more innovative for years. Management was particularly frustrated that the organization was unable to capitalize on the strong market position and large cash flow of its various late life cycle drugs. Instead, the R & D organization continued to focus on high-cost development of new chemical entities discovered in its own labs, and has mostly emerged empty-handed. While the late life cycle drugs continued to generate profits, management was worried that R & D would never come up with enough new compounds to replace its aging market leaders. Something new and different had to be done. After studying the problem, Gargantua management concluded that its sheer size was preventing the organization from tapping into innovative approaches. They wanted to see if a smaller, more nimble organization could succeed in growing a business where Gargantua had failed. Hence *Inventrix* was born.

Inventrix has a sizable workforce with a full complement of R & D, manufacturing, marketing and sales skills, but essentially no pipeline at all beyond the various product line extensions already in the works. Most of those line extensions are expected to produce only minor sales increases and will not do much to forestall the eventual decline of its marketed products. Current sales are $1 b, but are expected to tail off to $750 m in the next five years if no major new product is developed. In addition, Gargantua has given *Inventrix* a goal of doubling sales (to $2 b) during the same five-year time frame. Clearly, something completely new and different is required, and *Inventrix* is banking on a variety of strategic innovations to help it grow.

Inventrix management is keenly aware that trying to double sales over five years will require an incredible amount of energy, focus and coordination from the entire organization. They are also aware that the Gargantua culture has emphasized basic research in the R & D organization, but simultaneously encouraged short-term thinking everywhere else. Obviously, expectations will have to be set in all departments that encourage cooperation and teamwork and prevent people from working at cross purposes. A high level of energy will have to be maintained for five years, and that energy will have to be channeled into growing the business. In order to ensure that the organization keeps its focus and stays on track, *Inventrix* management wants to create a measurement system that encourages cooperation and teamwork, while simultaneously encouraging innovation and independent thinking – a tall order!

CROmed

CROmed is a contract research organization (CRO) that has been in business for ten years and employs 200 people. *CROmed* provides complete clinical trial services to pharmaceutical R & D organizations, but has distinguished itself because it has succeeded in cracking the enrollment problem. *CROmed* works in only one therapeutic area (TA), and has never missed an enrollment deadline. *CROmed* has conducted over 50 studies for its clients and enrollment for every one of them has come in on schedule – and most of them have come in ahead of schedule. *CROmed* has done this by cultivating a strong, loyal network of sites in its chosen therapeutic area, and creating a proprietary enrollment forecasting tool that has proven almost infallible.

Management has worked very hard at perfecting its enrollment scheme and now wants to increase sales. *CROmed*'s sales have been growing at about 10 percent per year, and are now at $30 m. Everyone feels as if the enrollment tools are now so well-honed that *CROmed* can expand beyond its current TA and tackle other areas. It seems perfectly reasonable to hit $50 m in sales in two years (a 30 percent annual growth rate). Everyone is excited about the opportunity.

There's only one problem: quality is horrendous. *CROmed* delivers great enrollment results, and monitoring is fine, but the data management, statistics and medical writing functions are slow and mistake-ridden. Meanwhile, *CROmed* is fielding a new electronic case report form (CRF) system that is causing great problems for the organization. Each new CRF takes an incredibly long time to create and the first release tends to have many errors. Ultimately, *CROmed* delivers a high-quality product to its pharma clients, but the problems in statistics, data management and medical writing mean that the reports have to be redone several times and are always late. Clients are so thrilled with the enrollment results that they are willing to put up with the other delays, but the costs of fixing all those mistakes is destroying *CROmed*'s bottom line (the cost of correcting the errors eats up all of its profit), and with so many quality problems it's getting harder to bring in new clients.

Management has decided that they need to get to the bottom of the quality problems (employees will tell you management has been saying this for years but has never acted!). There have been so many client complaints that this time the leadership team seems to be sitting up and taking notice. The chief operating officer (COO) has decided to review the strategic plan and develop a set of metrics that will get the entire organization immediately focused on quality. The chief executive officer (CEO) is very enthused about the idea, but is really focused on the 30 percent annual growth that has been agreed to with the board of directors. How to achieve the quality that's required to grow and still make their two-year growth plan is the real challenge for *CROmed*.

BioStart

BioStart is a 100-person biotechnology startup. It started with ten scientists who left when their company was taken over five years ago by Gargantua. They took with them some promising new ideas for cancer research that Gargantua dropped after the purchase.

They've been toiling away and are growing rapidly. Several venture capital firms have invested heavily in the promise of their new chemical entities (NCEs). If their compounds can get to market successfully, it looks like they could be blockbusters.

Pre-clinical and Phase I clinical trials are going well. *BioStart* is now getting ready to move into Phase II and is thinking about a combined Phase II/III trial on one of its compounds. But the growth in personnel and workload is taking its toll on focus and efficiency. When there were only 10–20 people in the organization, information was easy to come by in the way of a simple hallway conversation. Everyone ate together at lunch and each person was excited to share any news – whether good or bad – with his or her colleagues. But things are different with 100 people – and *BioStart* plans to hire another 50 people this year. The founders are much busier and less connected to day-to-day events, and they no longer know every employee well. The larger organization means that there are now separate functions, and information often gets bottled up within a functional group. Knowing what's going on has become more and more of a problem. Management is now worried that – with three to four clinical trials going on at once – it will be hard to ensure that everything is going well, that the various trials will stay on schedule and budget, and that problems are caught before they blow up. Everyone is all too aware of their 'big pharma' heritage in which small problems often didn't surface until they had become big problems, and most people were in full-time fire fighting mode. No one wants that to become the norm at *BioStart*.

To head off these problems as the organization grows, *BioStart* management has decided to build a simple clinical trial tracking system. They flirted with the idea of a complete balanced scorecard (BSC), but decided to start with a tracking system that would allow them to quickly obtain tracking data on their projects.

VirtuPharm

VirtuPharm is a mid-sized, US pharma with an established set of marketed compounds and a small but strong pipeline. *VirtuPharm* has always done its R & D in-house, with as-needed support from CROs. There has always been sufficient funding for the pipeline and relatively little time pressure, so *VirtuPharm* R & D has been able to operate in a fairly relaxed and casual environment.

But recently, two events have conspired to change that. First, *VirtuPharm* management has entered into strategic partnering talks with a foreign pharma that has several drugs they would like to introduce to the US market. These drugs would have to go through pivotal trials in the US, and some would have to go through some Phase I and Phase II testing. Second, *VirtuPharm*'s most profitable marketed drug has experienced some negative results in long-term Phase IV testing. While unlikely, there is a remote possibility that the FDA could ask *VirtuPharm* to take the drug off the market.

The fear of losing the market for their biggest selling product, combined with the sudden increase in their pipeline, has convinced *VirtuPharm* that they need to develop some long-term CRO relationships and abandon the ad hoc approach that R & D had been using.

Discussions with several CROs have been started, and the question of how to define a mutually beneficial relationship – and measure it – has arisen. *VirtuPharm* wants to find a way to create and measure true pharma–CRO partnerships.

CHAPTER 2

Selecting a Metrics System

I've worked with dozens of companies to build metrics systems, and it seems that every company wants to build a slightly different system. Here are just a few examples of metrics system goals:

1   Measure the performance of clinical trials
2   Make sure each clinical trial is completed on schedule and on budget
3   Improve cycle times of clinical operations
4   Write and execute better protocols
5   Make our clinical organization more efficient and effective
6   Measure the performance of our CRO contractors
7   Create lasting pharma–CRO relationships
8   Bring drugs to market faster
9   Make clinical teams more effective
10   Increase innovation in R & D.

On one level, the goals of all these systems are the same: do things better, more efficiently and faster. However, when it comes to developing the measurements, each system is slightly different.

IDENTIFYING METRICS SYSTEM GOALS

Let's look at some examples of metrics system goals. The first two systems listed above ('Measure the performance of clinical trials' and 'Make sure each clinical trial is completed on schedule and on budget') might appear at first glance to be the same. In reality, they are very different.

System 1 seeks to measure with no specific goal in mind, while System 2 is attempting to make actual performance meet or exceed forecast. Thus System 1 could look at any number of possible metrics (cycle time, quality, cost, labor, error rates), while System 2 focuses on the actual milestone and cost achievement vs. the forecast. Those building System 1 will have an unlimited number of measures to choose from, while those building System 2 will have only a few. At the same time, those building System 1 will have no idea which metrics might be most useful, since the system has no clear purpose. Figure 2.1(a) compares these two systems.

This example illustrates the first principle in selecting and developing a metrics system:

Identify the goal (or purpose) for which you are designing your metrics system.

Since System 2 has a clear goal, there is a reasonable chance that the organization *will* pay attention to the results and use them to achieve the goal of on-time/on-budget performance. System 1, however, probably won't be around for long (or perhaps it will stay around for quite a long time, and data will be collected and filed, but never used!). It has no clear purpose and no one will know how to use the results. Only people who take the time to really analyze the results will have any use for the system, and such people are rare in the frenzied world of pharmaceutical R & D.

Many times I have been invited to review an existing metrics system, in some cases because an executive is frustrated with the resources being expended on collecting and reporting metrics. On several occasions I have encountered metrics systems that are highly evolved and require support groups of half a dozen or more people to feed and care for the system. In one instance, I encountered a group of eight people who produced beautiful, 150-page monthly reports filled with full-color charts and graphs on virtually every R & D operational metric you could imagine. But the staff complained that no one really seemed to use – or even care about – their metrics reports, while management complained about the cost. Once I had interviewed the staff as well as recipients of the monthly report (which included several senior executives), it became obvious that there was no real focus to the system. It was simply a collection of performance metrics that had developed over time. Various recipients would use one metric or another for their own purposes, but no one systematically reviewed the entire set. The metrics system had become an end in itself. Clearly there were wonderful gems buried in the data, but no one was actually mining and using that information. In fact, less than 20 percent of the metrics were used to manage and improve the business. This is a prime example of the kind of problems associated with systems that don't have a clear goal or purpose, such as System 1.

GAINING COMMITMENT TO YOUR METRICS SYSTEM

The second and third systems ('Make sure each clinical trial is completed on schedule and on budget' and 'Improve cycle times of clinical operations') are also different, as shown in Figure 2.1(b). System 2 looks at each trial individually and attempts to achieve success. When the team finishes the trial, it moves on to another one. Any lessons learned are carried along by the team members themselves. There is no attempt at continuous learning and improvement, or even communicating the lessons learned to other teams. System 3, on the other hand, looks at a variety of trials and attempts to make improvement over time and across trials. This system is focused, in that it looks at improving cycle times, but doesn't attempt to make changes, corrections or improvements within any single trial, as does System 2. Rather, it seeks to make longer-term improvements to the way work is accomplished in the organization. This is a much bigger task and requires more time and resources. However, System 3 has the potential to influence the entire organization and make fundamental change, while System 2 affects only a single trial or team and probably

Attribute	System 1: 'Measure the performance of clinical trials'	System 2: 'Make sure each clinical trail is completed on schedule and on budget'
Goal	Unknown	Assure actuals match forecast
Focus	All trials	Single trial
Range of available metrics	Large – any clinical operations measure	Small – only those which measure actual vs. forecast milestone and cost performance
Ability to align metrics with goal	Impossible	Good
Potential effectiveness of system in improving performance	Ineffective	Effective

Figure 2.1a   Comparing unfocused vs. focused metrics systems

Attribute	System 2: 'Make sure each clinical trial is completed on schedule and on budget'	System 3: 'Improve cycle times of clinical operations'
Goal	Assure actuals match forecast	Improve operating performance over time
Focus	Single trial	All trials
Range of available metrics	Small – only those which measure actual vs. forecast milestone and cost performance	Large – any measure which affects clinical operations performance
Ability to align metrics with goal	Good	Good
Potential effectiveness of system in improving performance	Effective on individual trials only	Effective over many trials in improving overall operating performance
Time required to collect sufficient data for decision making	Weeks	Years

Figure 2.1b   Comparing trial-by-trial vs. systemic improvement metrics systems

Attribute	System 3: 'Improve cycle times of clinical operations'	System 5: 'Make our clinical organization more efficient and effective'
Goal	Improve operating performance over time	Improve organizational performance over time
Focus	All trials	Entire organization
Range of available metrics	Large – any measure which affects clinical operations performance	Very large – any measure which affects operations or organizational performance
Ability to align metrics with goal	Good	Good
Potential effectiveness of system in improving performance	Effective in improving overall operating performance	Effective in improving overall organizational performance
Time required to collect sufficient data for decision making	Years	Years

Figure 2.1c   Comparing process improvement vs. organizational improvement metrics systems

achieves no long-lasting effect. On the other hand, System 2 can be used by any individual team and will yield results in the course of a few weeks or months, while System 3 must be adopted by the entire organization and data must be collected over a long period of time – perhaps years – before there is enough information to support change and improvement.

This illustrates the second principle in selecting and developing a metrics system:

Make sure the organization is committed to sustaining the metrics system you choose.

If you attempt to implement System 3 with a goal of improving the entire clinical operations process, you will need data from every clinical trial team – and hence the cooperation of every team. Any team that doesn't participate and provide data weakens the overall system in the sense that there is less data to analyze and examples of best practices may be missed. On the other hand, if you implement System 2, each team that decides to participate will get its own feedback and results and can take appropriate actions. Teams that choose not to participate do not negatively affect the metrics system (although they will likely run into operational problems that the metrics system would have detected).

GAINING MANAGEMENT SUPPORT

Now let's look at the third and fifth metrics systems ('Improve cycle times of clinical operations' and 'Make our clinical organization more efficient and effective'). System 3 seeks to improve the processes that are used in performing clinical trials. On the other hand, System 5 seeks to improve the entire organization and the way it functions. While System 3 looks at cycle times, quality and the like, System 5 looks at those metrics plus others: organizational performance, team performance, employee satisfaction, customer satisfaction, and more. If System 3 struck you as a significant leap in complexity over System 2, System 5 represents another significant leap in complexity. In System 5, metrics must be chosen that measure not only processes, but also the performance of individuals, teams, and the organization as a whole. We also need more sophisticated measurement tools and more labor to gather and analyze the data, since we now have customer and employee satisfaction surveys and a variety of other qualitative measurements. Of course the payoff from System 5 is significantly larger as well (see Figure 2.2). Not only can we improve processes, but we can make sure that the work we are doing is more focused on what our customers want. Remember the discussion of effective work in Chapter 1? By focusing on what our customers want, we are by definition increasing our effective work (more about who R & D's customers are and what they want is discussed in Chapter 3). Not only do cycle times get faster, but the percentage of effective work increases; we've moved from doing everything faster to doing the most effective tasks faster (and hopefully eliminating the ineffective tasks).

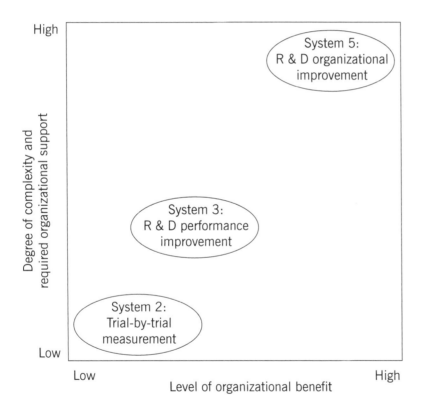

Figure 2.2 Different levels of management support and system complexity are required to achieve different levels of organizational improvement

This comparison illustrates yet another principle of selecting and implementing metrics:

Make sure you have sufficient management support at the appropriate levels.

Metrics System 2 requires little or no management support. Any team can choose to use or ignore the system. System 3 requires fairly strong support from senior clinical operations leadership. All teams must contribute data if the system is to prove useful, and the data must be both correct and consistent across all teams (both of which are harder to do than you might think!). System 5 requires far more commitment, and from an even higher level. Senior R & D executives must actively support the system, as must human resource executives and customer executives (marketing, sales and so on).

Figure 2.3 looks at each of the ten example systems listed previously and describes the advantages and disadvantages of each.

THE METRICS SYSTEM QUESTIONNAIRE

Before we continue you may find it helpful to use the questionnaire in Figure 2.4. It's short (only six questions), but it will help you figure out whether you've got a good foundation for your metrics system. We'll use the results in the next section, so take a few minutes to jot down your answers. As you progress through the book, you may want to come back to the questionnaire and re-evaluate your answers. That's fine. The goal is to make sure you end up with the right metrics system for your needs.

THREE CLASSES OF METRICS SYSTEMS

Take another look at Figure 2.3. You'll notice that there's a 'metrics system class' listed for each of the ten sample systems. As the figure implies, we can divide pharma R & D metrics systems into three broad classes:

1 Tracking
2 Performance improvement
3 Organizational improvement.

Once you've decided to build your metrics system, you need to figure out which class of system is most appropriate for you. That's important because the specific metrics you choose to include in your system depend on the class you choose. We'll look briefly at each system and then discuss some examples. Figure 2.5 shows the attributes of each system and Figure 2.6 graphically illustrates how the three classes of metrics systems relate to each other.

Tracking systems

Tracking systems are the most basic kind of pharma R & D metrics system. Their goal is simply to alert a clinical trial team or a project team of evolving problems that might endanger the schedule or budget (or both). Problems that these systems typically attempt to address are:

Metrics system goal	Does system have a related improvement goal?	Can goal be quantified?	Scope/focus of system	Metrics system class
1 Measure the performance of clinical trials	No	No	Clinical operations processes	Tracking
2 Make sure each clinical trial is completed on schedule and on budget	Yes	Yes	Clinical trial team	Tracking
3 Improve cycle times of clinical operations	Yes	Yes	Clinical operations processes	Performance improvement
4 Write and execute better protocols	Yes	Possibly	Medical/clinical operations	Performance improvement
5 Make our clinical organization more efficient and effective	Yes	Yes	Clinical operations organization	Organizational improvement
6 Measure the performance of our CRO contractors	Yes	No	Contracts/CROs	Tracking
7 Create lasting pharma–CRO relationships	Yes	Possibly	Clinical operations organization/CROs	Organizational improvement
8 Bring drugs to market faster	Yes	Yes	R & D processes	Performance improvement
9 Make clinical teams more effective	Yes	Possibly	Clinical trial team	Organizational improvement
10 Increase innovation in R & D	Yes	Yes	R & D organization	Organizational improvement

Figure 2.3   Attributes of various metrics systems

SELECTING A
METRICS
SYSTEM

Goals:

1 Is your metrics system focused on attaining a particular goal (examples of goals are 'Increase efficiency in clinical operations', 'Submit more NDAs per year', 'Decrease operating costs', 'Increase innovation across the organization')?

   (a)   If 'Yes', give yourself 5 points, *or*

   (b)   If 'No' or 'Not sure', give yourself 0 points: ____

2 Is your goal written down?

   (a)   If 'Yes' and you developed the goal yourself, give yourself 5 points, *or*

   (b)   If 'Yes' and it comes from your organization's senior leadership (or you are your organization's senior executive), give yourself 10 points, *or*

   (c)   If 'No' or 'Not sure', give yourself 0 points: ____

3 Is your goal quantifiable, that is, can it be measured easily (examples of quantifiable goals are 'Increase efficiency in clinical operations *by 15%*', 'Submit *at least three* NDAs per year', 'Decrease operating costs *by 10% per year*', 'Increase innovation across the organization *by implementing two major innovations per year, where a major innovation will increase company profit by $1 m per year*')

   (a)   If 'Yes', give yourself 10 points, *or*

   (b)   If 'No' or 'Not sure', give yourself 0 points: ____

Management Support:

4 Has the senior leader or management team in your organization endorsed this metrics system?

   (a)   If you or another employee went to senior management with the idea *and* they endorsed it, give yourself 5 points

   (b)   If the organization's senior executive initiated the idea and came to you about it, give yourself 10 points

   (c)   If in addition to (a) or (b), the organization's senior executive has *publicly announced* his or her commitment to build this metrics system, give yourself 20 points

   (d)   Otherwise, give yourself 0 points: ____

5 Sometimes the metrics system requires obtaining data from other groups outside your organization (and who report up through separate chains of command), such as human resources (HR) or the finance group. For instance, you may want to build a metrics system that measures employee satisfaction, so

you'll need help from HR. Or you may want to compile cost data, so you'll need help from the financial group or controller's office. List all of these groups that must support you, *then*

- (a) Give yourself 5 points for each group that has said they will support you
- (b) Give yourself 0 points for each group that has said no or has ignored your request for help
- (c) If you don't require data from any outside sources, give yourself 10 points: ____

Organization Support:

6 Ask five of your colleagues around the organization whether they think the metrics system you are building will be useful or not. Focus on talking to people who will be key users of your data (for example, department managers, executives, team leaders)

- (a) Give yourself 5 points for each colleague that is enthusiastic about the project
- (b) Give yourself 2 points for each colleague who is neutral but persuadable
- (c) Give yourself 0 points for each colleague who is negative: ____

Add up your points and put the total here: ____

Scoring:

>80 points: You're in good shape. It looks like you've got plenty of focus and support.

>60 points: You've got work to do, but you should succeed in developing your system.

>40 points: You've got some selling to do before you can be sure of success.

>20 points: You're just beginning to develop the basis of your metrics system. Increase your score to at least 40 before you attempt to implement the system.

<20 points: Keep reading, but you've got work to do before you embark on a metrics development project.

Figure 2.4   The pharma metrics questionnaire

Class	Tracking	Performance improvement	Organizational improvement
Focus	Maintain project schedule and cost	Improving operational performance over time	Improving all aspects of organizational and operational performance
Goal	Provide early warning of project problems	Track and continually improve processes and operational performance	Track and improve performance of the entire organization
Typical metrics	Predictive project metrics	Predictive and retrospective project and process metrics	Predictive and retrospective financial, satisfaction, organizational growth and performance metrics
Advantages	Helps prevent problems on current project	Increases overall operational performance	Increases overall organizational health and improvement. Aligns all departments and employees around key goals
Disadvantages	Doesn't help with future projects. Doesn't catch systemic errors	Requires multiple projects to develop trends and gather sufficient data	Requires multiple, cascading measures to be successful. Requires highly disciplined tracking and follow-through
Management commitment level	Low	High	Very high

Figure 2.5   Attributes of metrics systems classes

- schedule and cycle time problems such as late/slow site initiation and enrollment, or slow data collection and database lock
- quality problems such as excessive queries or final report problems
- cost problems such as excessive screen failures or costly amendments.

An organization that uses a tracking metrics system intends to look at each project or trial individually and has no interest in using the metrics system to help systematically improve performance over time. To some, using a tracking system may seem pointless. Why go to all the trouble of building a metrics system if you won't be able to use it to continually improve your operations? But tracking systems can be very useful in some situations:

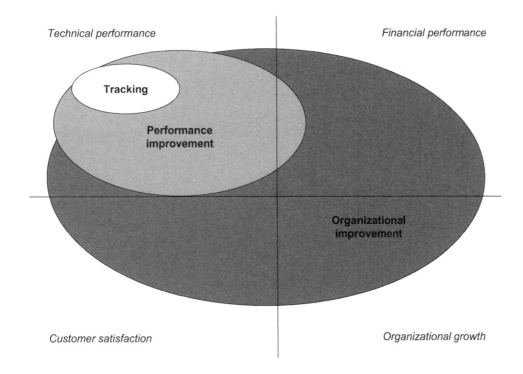

Technical performance

Financial performance

Tracking

Performance
improvement

Organizational
improvement

Customer satisfaction

Organizational growth

Figure 2.6 Relationship between the three metrics system classes

- The system needs to be easy to develop and implement.
- Buy-in from the entire organization is not desired or available.
- Costs to build and maintain must be kept to a minimum.
- The organization is too small to have the infrastructure needed for a larger system.

In fact, one of the most common failures of more ambitious metrics systems is that they don't have enough buy-in from the organization to be sustained (see the pharma metrics questionnaire in Figure 2.4). Without sufficient buy-in, the various parts of the organization fail to provide the data needed to build and maintain the database and it becomes impossible to produce any useful trend data. In those situations, you may find that it's best to start small by setting up a tracking system. Once the tracking system has proved its value, you'll have executives who were originally cool (or even hostile) to the idea of a metrics system come banging on your door for more sophisticated data. So, if you find yourself with a relatively low score on the questionnaire, you may want to start with a tracking system until you find you have raised your score a bit.

Performance improvement systems

The next class above tracking systems are performance improvement systems. Instead of measuring each project separately, a performance improvement system measures all projects in the organization. The goal is to create a system that can measure trends across time and provide clues as to how to improve the processes for getting work done. Measurement can be at the clinical trial level, the compound level or the functional level.

At the clinical trial level, the metrics themselves might look very much like those used for a tracking system. However, the tracking system goal of achieving on-schedule/on-budget performance is only one aspect of the performance improvement system. Performance improvement systems are much more sophisticated than tracking systems because data comes in from all over the organization and must be processed and validated. A database manager is typically required to make sure the data is entered in a timely manner and it is accurate and consistent across the organization. Those that are tardy in entering data need to be reminded, and management needs to be insistent on getting the data.

On the output side, periodic reports must be generated and distributed, and care needs to be taken to ensure that the data is objective and not confrontational or accusatory. When developed and used properly, these systems have the ability to pinpoint areas where improvement could yield faster cycle times, better quality, higher efficiency, lower cost and more 'effective' work.

Organizational improvement systems

Finally, there are organizational improvement systems. This class of system attempts to measure and track the performance of the entire organization. While performance improvement systems focus on processes and technical performance, organizational improvement systems include:

- technical performance
- financial performance
- employee performance and satisfaction
- organizational leadership, growth and maturity
- customer satisfaction.

Organizational improvement systems by definition include performance improvement systems. The Malcolm Baldrige National Quality Award (MBNQA) introduced the concept of organizational improvement when it was created in the US in 1988,[1] and Robert Kaplan and David Norton popularized this class of metrics systems with their first balanced scorecard article in 1992.[2] The MBNQA goals are typical of the goals for an organizational improvement system:

- *'Delivery of ever-improving value to customers, contributing to marketplace success*
- *Improvement of overall organizational effectiveness and capabilities*
- *Organizational and personal learning'*.[3]

Organizational improvement systems are the most complex class of metrics systems to develop and administer. They require input over a long period of time both from within and outside the organization (clinical teams, finance, human resources, customers, suppliers, and so on) and use a highly refined set of measures. On the other hand, they

---

1   See http://baldrige.nist.gov for an overview and history of the Malcolm Baldrige National Quality Award.
2   'The Balanced Scorecard – Measures that Drive Performance', *Harvard Business Review*, Jan–Feb 1992.
3   '2004 Criteria for Performance Excellence', Baldrige National Quality Program, National Institute of Standards, p. 1
    (http://baldrige.nist.gov/Business_Criteria.htm).

have the ability to drive tremendous improvement and change across the organization. In addition, they have the unique ability to help align all parts of the organization around a few key goals. For instance, a major threat to CRO profits can come from employee turnover. Loss of several valued employees can frustrate pharma customers, create dissatisfaction among CRO employees and cost the CRO significant time and money to train replacements. Thus employee turnover must be measured and reduced. Turnover would never be included in a performance improvement system because it's not part of any process. However, it is a standard measure used by MBNQA organizations and balanced scorecard systems.

CASE STUDIES

Let's look at our example companies to see which class of system each should use.

Inventrix has a goal of doubling sales in five years through innovation. Management needs to create a sense of shared destiny among its employees and an intense focus on the goal. Working on small, short-term gains will never be successful, so everyone in the organization needs to learn to patiently but persistently strive for long-term growth. *Inventrix* management wants to create a measurement system that encourages cooperation and teamwork, while simultaneously fostering innovation and independent thinking. Which class of metrics system is most appropriate here? If you said organizational improvement, you're exactly right. Short-term tracking systems will have little value to *Inventrix*, since they look only at single projects. Performance improvement systems will help in some respects, but they won't help with the teamwork, cooperation and innovation issues. Only an organizational improvement system will help here.

CROmed is trying drastically to increase quality. Customer complaints about quality are significant and growing. Employees are increasingly frustrated. There is concern that the 30 percent annual sales growth rate could be endangered by poor quality. What class of metrics system is appropriate here? This situation is a bit more ambiguous than *Inventrix*. Ideally, an organizational improvement system should be instituted. Issues such as customer satisfaction, employee satisfaction and growth, and financial performance are all captured in an organizational improvement system. On the other hand, the core of *CROmed*'s problem is likely to be process quality, which is at the center of a performance improvement system. Often, organizations like *CROmed* start with a performance improvement system and develop it into a full organizational improvement system at some point in the future. Since performance improvement systems are simpler and faster to develop, this can sometimes be the quickest way to get results. In this approach, *CROmed* would first build measures of process quality and independently carry out some customer satisfaction surveys to determine the primary causes of customer dissatisfaction. They would also field a simple employee satisfaction questionnaire to give employees a medium in which to surface internal quality problems. That would lead to development of the performance improvement system metrics. Once improvements begin to take hold, *CROmed* could then institute the other types of metrics appropriate for an organization improvement system.

BioStart is facing significant growing pains. The rapid growth in staff, coupled with the cost and schedule pressures imposed by venture funding and the relatively large number of later-phase trials are unnerving for such a small organization. An organizational improvement system might be the obvious choice for *BioStart*, since it deals with both the performance aspects and the organizational aspects of the firm. However, the commitments required of such a system make it difficult to implement under *BioStart's* circumstances. Because the immediate focus is on clinical trial success, the simplest approach here is to start with a tracking system to ensure that the goal of on-time, on-budget trials is met. If *BioStart* management can be sure that each trial is moving ahead in an expeditious and cost-effective manner, they won't be tempted to micro-manage the projects. They can then focus their energies on the longer-term issues that are dealt with in performance improvement and organizational improvement systems. So the best answer here is to start with the near-term problems via a tracking system.

VirtuPharm is looking to build long-term, cross-organizational relationships. Tracking systems focus neither on the long term nor on relationships, so they are out of the running in this case. Performance improvement systems do allow cross-organizational focus, but only at a process level (for example, measuring how efficiently the partnership gets work done). They don't measure real organizational issues (mutual satisfaction of the parties, cross-organizational teamwork and so on). If *VirtuPharm* really intends to measure their relationships with their partners, only an organizational improvement system will work.

SUMMARY

Picking the right metrics system is really defined by your measurement goal. As we saw in our discussion of the example companies, most organizations can use one of several types of metrics systems. There are three principles for successful selection of a system:

1   Identify the goal (or purpose) for which you are designing your metrics system.
2   Make sure the organization is committed to sustaining the metrics system you choose.
3   Make sure you have sufficient management support at the appropriate levels.

Building a metrics system without both of these principles firmly in hand will make your life very difficult indeed!

Use Figure 2.3 to help you select the best metrics system and goal for your application. Also, use the metrics system questionnaire in Figure 2.4 to help you figure out whether you've got a sufficient system definition and enough organizational support to be successful.

Once you've decided on your goals and acquired support, decide whether you want to build a tracking system, performance improvement system or organizational improvement system. The four company examples above should help you decide on the best system for your purposes. Since these systems require time, energy and commitment to build, don't bite off more than you can chew. Use the system that fits your needs with the minimum number of bells and whistles.

CHAPTER 3

Creating the Foundation Using Strategy Maps

I was planning a short vacation with my wife recently. We wanted to visit our daughter at college in Boston. I had blocked out a week on my schedule, so we also had several days in which we could relax and unwind – just the two of us. We wanted to consider all our options, so we started talking about seeing a show in New York, driving through New England to 'see the colors' (it was October and the trees were all turning beautiful shades of yellow, red and orange), visiting Maine (where I had never been) or Quebec (where she had never been), visiting friends in Vermont or upstate New York ... Before we knew it, we had dozens of options. We tried to whittle down the list. We looked at flights to see if we could make some distinctions by minimizing the cost and the travel time. We looked at rental cars, driving distances and hotels to see if we could make distinctions by minimizing the cost and driving distance, or maximizing our comfort. We ended up with many different metrics, mounds of data and multiple tradeoffs: it's cheaper to get to Quebec than New York, but we can't get there directly; Maine will be beautiful, but we'd like to do cultural things as well as relaxing in the woods. The more we worked at the problem, the messier it got. The whole planning process was giving us headaches. Planning a relaxing vacation just should not have been this difficult!

Finally, it occurred to us that we'd forgotten our goal: to visit our daughter and have some time to relax and unwind. As soon as we refocused on the goal, the problem got simple: spend as little time as possible traveling and as much time as possible relaxing. The solution then became obvious: fly to Boston – where we could relax and get a cultural 'fix' – and take a day trip to see the colors; invite our friends to meet us in or near Boston; and of course see our daughter whenever her school schedule would permit. All of the complicated tradeoffs we were trying to make became moot when we put them in the context of our goal.

LINKING ACTIONS TO GOALS

Perhaps you've suffered one of these unfocused episodes, only to eventually realize that your problem was that you'd strayed from your original goal. We've all had personal experiences like that. As it turns out, organizations suffer from the same lack of focus – and it happens a lot. Sometimes this happens as committees or subgroups discuss options and drift towards alternatives. Sometimes it happens when a senior executive issues a directive. There's nothing like a 'take that hill!' command from a senior executive to send an entire organization into a fit of unfocused exuberance: everyone in the organization running off to do whatever she or he can to make the boss's wish come true. In order to avoid these unfocused episodes, it is critical to keep the overall goal in mind as organizations make decisions.

Perhaps you've heard of the Iditarod. It's a ten-day, 1000-mile sled dog race from Anchorage to Nome, Alaska. The dog teams for this race obviously must have both speed and incredible stamina to win the race. But before a musher (the person who runs the dog sled) can worry about endurance, he or she has to make sure that all the dogs are pulling in the same direction. The organizational unfocused episodes we just talked about are a little like a musher who hooks each of the dogs to a different part of the sled and then yells 'mush!' The poor dogs pull with all their might, but the sled goes nowhere – or maybe gets ripped to pieces! Like the sled dogs, we can pull as hard as we want, but if everyone is pulling in a different direction, the organization goes nowhere. In the worst of circumstances, people start pointing fingers at each other, accusing each other and avoiding responsibility, thereby ripping the organizational sled apart. In order to avoid all this, all we have to do is get everyone pulling in the same direction so that we harness the energy of the group and begin making progress.

Whether you're racing in the Iditarod, planning a vacation, or trying to achieve organizational goals, the bottom line is the same: you have to link your actions directly to

Figure 3.1a   When the sled dogs are attached to different parts of the sled, they all pull in different directions, ripping the sled apart!

Figure 3.1b When the sled dogs are all aligned and pulling in the same direction, progress can be swift

your goals. Otherwise, you'll end up with lots of wasted (that is, 'ineffective') – or even destructive – work. Figure 3.2 shows the way to link actions to goals in an organization.

Let's take a look at how this works. First, it's critical to establish your goals. Are you trying to develop more drugs, develop them faster, develop them less expensively, or all of the above? You can imagine that a goal of developing all drugs faster might lead to a very different strategy than a goal of developing only a few blockbuster drugs. The first goal

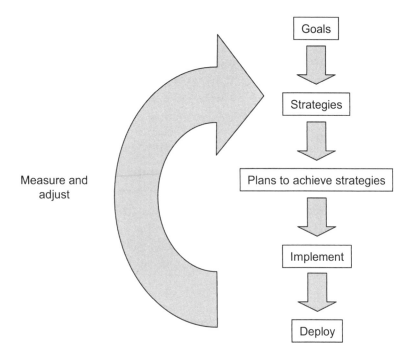

Figure 3.2 Linking actions to your goals provides organizational focus and alignment

pushes you to focus on the process of turning each new compound into a marketable drug (regardless of its market potential). It doesn't matter whether the drug is a multi-billion dollar blockbuster, an orphan drug that will help a small population with no other available cure, or a drug that has little chance of ever making it to market at all due to a potential lack of efficacy. Regardless of the drug, a goal of developing all drugs faster will cause the organization to focus on reducing cycle time.

The second goal, on the other hand – developing only those drugs that are likely to become blockbusters – pushes the organization to identify those compounds that will not become blockbusters and weed them out quickly. The emphasis is on the selection process rather than the development process. With a blockbuster drug goal, the organization may decide that a fast development cycle – while very desirable – is much less important than the preceding discovery and weed-out processes and therefore might decide to ignore cycle times in the face of limited resources.

If the organization chooses a goal that is greater than its reach, then it risks getting nothing at all accomplished. For instance, if a pharma chooses to do both of the above goals – improve development cycle times *and* focus on blockbusters – then it will have to deal with potential resource problems during its strategy development, or face the problem of insufficient resources to do either well. Unless the organization is careful, it runs the risk of attaching its dogs to different parts of the sled.

Once you've established your goals, you can define strategies that will help you achieve them. These strategies tend to be fairly broad, high-level concepts such as improving development cycle times by 50 percent or creating a process for effectively selecting drugs for development. We'll explore a method for developing these strategies in the next section on 'strategy maps'. Each strategy is linked to the goal that it supports.

Finally, the strategies are extended to specific actions and plans for accomplishing them (see Figure 3.2). In contrast to the broader strategies, which are defined by the organization in light of organizational goals and must be executed by the entire organization, detailed actions can actually be executed by individual members or small groups within the organization.

Once the goals, strategies and actions have been developed, the metrics system (which, after all, is the point of this book) can be used to link all of the pieces together and ensure that all parts of the organization are aligned around the same goals, strategies and actions. In effect, the metrics system is like a 'sled pull meter' that allows the Iditarod musher (or organizational leader) to figure out whether all the dogs are pulling consistently in the same direction, and to confirm that the direction in which all the dogs are pulling is the correct one to win the race.

Now, you might think that this is overkill. All we need to do is look where we're going and then look at what the dogs are doing, right? Well, it turns out that even in sled dog racing there are a variety of strategies that can be used. Winning the race actually starts years before, with dog training and endurance building. Most mushers make sure all of their dogs have an acceptable level of strength and endurance, and then train their strongest dogs to be superstars. The hope is that the strongest dogs will provide examples for the weaker ones and spur them on to better performance. In the 2004 Iditarod, Mitch Seavey won a particularly competitive race, by using a different strategy. He attributed his win to making the weakest dogs as strong as the others rather than making superstars of the strongest dogs. Daily performance metrics for each of the dogs helped him create an invincible team. Working the team to make sure even the weakest were up to the task gave his dog team the endurance required to cover 1000 miles in record time. He emphasized a teamwork strategy more than a superstar strategy. The same can be said for any organization. Once you know where you're going (your goal), you have to create a plan for getting there (your strategies), detailed plans (your actions) and a set of metrics that tell you whether the plans and actions are being executed effectively.

Organizational improvement is much more complex than sled dog racing, since we usually can't 'see' the goal and we can't keep daily tabs on the thousands of people who we've asked to help achieve that goal. Conversely, an individual employee down in the trenches often can't relate to ('see') a top-level goal in the organization. For instance, a bench chemist doing day-to-day high performance liquid chromatography (HPLC) analyses has trouble relating this work to a corporate goal of increasing profit or reducing time to market. She is simply doing what she was hired to do. However, if we break down the corporate goals and give that chemist a goal of increasing profit or reducing time to market by making sure that every analysis is done accurately and to customer

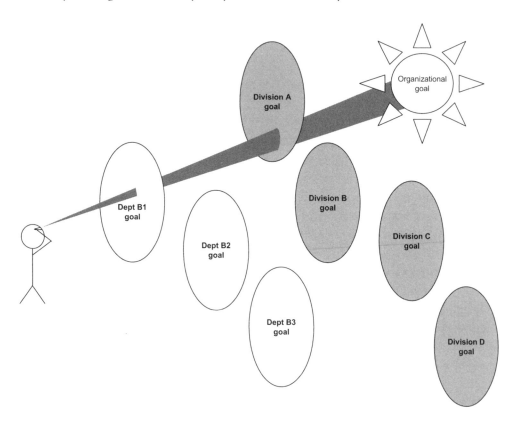

Figure 3.3   In order to create a line of sight from an individual in a department to an organizational goal, the individual must be able to 'see through' the various organizational layers

specifications each and every time, she can easily 'see' the goal. So we need a way of creating a 'line of sight' from each person's individual actions to the organization's overall goal. A good metrics system is the vehicle that provides that line of sight (see Figure 3.3).

STRATEGY MAPPING

So, how do we create this linked set of goals, strategies, action plans and metrics? There have been lots of methods proposed over the years. Recently, Robert Kaplan and David Norton proposed a method they have dubbed 'strategy mapping'.[1] It's a robust technique for developing strategies and action plans that are clearly linked to an organization's goals. Kaplan and Norton give a detailed treatise on strategy maps in their book *The Strategy-Focused Organization*'.[2] While there are many nuances to the technique, the basics can be summarized as follows:

1   First, identify your goal.
2   Determine specific *financial strategies* that must be implemented to achieve the goal.
3   Determine the *customer strategies* that must be implemented to achieve the financial strategies and delight customers.
4   Determine the *performance improvement strategies* that we must implement in order to create delighted customers.
5   Determine the *organizational growth strategies* that we must achieve in order to improve performance.

Figure 3.4 illustrates the strategy map development process. In essence, the strategy map process helps you move successively from one layer of strategy to the next, thereby creating a consistent set of strategies.

You may wonder at this point why the four strategy areas of financial, customer, performance improvement and organizational growth are selected for mapping. Here's the way to think about it. As scientists, we tend to focus on the technical areas of the business. We concern ourselves with doing things more accurately and effectively. Hence, creating strategies that improve performance seems reasonable and natural. It's easy to think about doing things faster or with higher quality. However, if we only look at the world from that one perspective, we run the risk of financial bankruptcy. How many Internet firms have pursued some new technology and have gone out of business because they paid no attention to making money – or even breaking even. If you focus exclusively on making a better product, more often than not you'll eventually run out of money. So it's important to think about both improved performance and financial realities such as profit. But even those two aren't sufficient. You can improve performance and keep costs down, but if you don't produce something that your customer wants, no one will buy it. So getting to market with a product that no one wants isn't much better than breaking the bank during development. Finally, you can improve performance, create financial stability

---

1.   'Having Trouble with Your Strategy? Then Map It', *Harvard Business Review*, Robert S. Kaplan and David P. Norton, Sept–Oct 2000.
2   *The Strategy-Focused Organization: How Balanced Scorecard Companies Thrive in the New Business Environment*. Robert S. Kaplan and David P. Norton. Harvard Business School Press.

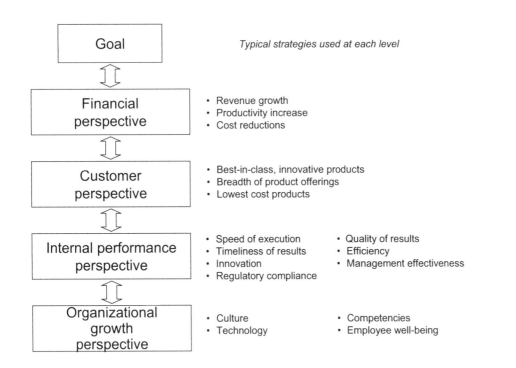

Goal

Financial
perspective

- Revenue growth
- Productivity increase
- Cost reductions

Customer
perspective

- Best-in-class, innovative products
- Breadth of product offerings
- Lowest cost products

Internal performance
perspective

- Speed of execution
- Timeliness of results
- Innovation
- Regulatory compliance
- Quality of results
- Efficiency
- Management effectiveness

Organizational
growth
perspective

- Culture
- Technology
- Competencies
- Employee well-being

Typical strategies used at each level

Figure 3.4 The strategy map development process. (Source: 'Having Trouble with Your Strategy? Then Map It', R.S. Kaplan and D.P. Norton, *Harvard Business Review*, Sep/Oct 2000)

Strategy perspective	Example questions to ask as you build your R & D strategies
Financial perspective	To achieve our vision and goal, what financial strategies must we execute? Should we focus on revenue growth, efficiency or cost reduction?
Customer perspective	To achieve our vision, goals and financial strategies, what customer strategies must we execute? Should we differentiate ourselves through best-in-class products, breadth of our product offerings, or low-cost products?
Internal performance perspective	To achieve our vision, our financial strategies and our customer strategies, how must we improve our internal operations and performance? How do we enhance speed, timeliness, quality and efficiency of product development and delivery? How do we increase innovation? How do we meet or exceed regulatory requirements? How do we improve the way we work with CROs and suppliers?
Organizational growth perspective	To achieve our vision and our financial, customer and internal performance strategies, how must our organization grow and improve? What should our culture look like? What core competencies must we have or develop? What technologies must we emphasize? How should our departments and divisions interact and work together? How do we assure employee well-being?

Figure 3.5 Questions that will help you build your strategy map. (Source: 'Having Trouble with Your Strategy? Then Map It', R.S. Kaplan and D.P. Norton, *Harvard Business Review*, Sep/Oct 2000)

and satisfy customers, but you'll still fail if your organization isn't growing and your employees are too dissatisfied to stay with you. Your organization and employees are the engine that drives new products and ideas, so they are every bit as critical as the other three. Hence, any successful strategy has to have components in all four areas.

Some of the key aspects of each perspective are shown in Figure 3.5.

CASE STUDIES

We'll use our four sample companies to illustrate how strategy maps can be used in pharmaceutical R & D. In each case, I've provided one possible set of strategies for the organization. You might perceive that a very different strategy might be more appropriate. That's fine, especially since these companies are figments of my imagination anyway!

Inventrix

Inventrix has the following characteristics:

- The company has staked its future on strategic innovation.
- It must double sales in five years to $2 b while reversing the slide in sales from its current stable of drugs.
- Doubling sales over five years will require an incredible amount of energy, focus and coordination from the entire organization.
- The current culture emphasizes basic research in the R & D organization, but short-term thinking everywhere else. Cooperation and teamwork will be required to prevent people from working at cross purposes.
- A high level of energy will have to be maintained for five years, and that energy will have to be channeled into growing the business.
- Senior *Inventrix* management is relying on R & D to come up with new ideas that can be parlayed into breakthrough products.

A strategy map for *Inventrix* R & D might look like Figure 3.6. The first thing you'll notice is the goal. It doesn't say 'double sales' because that isn't solely R & D's responsibility – nor is it within R & D capability to achieve by itself. R & D will have to work with marketing, sales, manufacturing and other areas of *Inventrix*. An organizational goal such as 'double sales' can only be achieved with cooperation from other groups in *Inventrix*. R & D's goal must be something the group can achieve independently. This distinction may seem obvious, but it's critical: the goal at the top of a strategy map must be something that the organization (or subgroup of the organization for whom the map is developed) can achieve on its own, or with resources that it can control (such as suppliers or subcontractors). That way, the group has the ability to develop strategies that can actually be implemented! If the goal involves other players that R & D can't control (such as other divisions or partner companies), then the R & D organization can't be sure that its strategies will lead to successful goal attainment. In the case of *Inventrix*, R & D can control its ability to innovate, so the goal – while perhaps hard to measure – is attainable.

R & D management decided that there are two possible financial pathways available:

1   Create new products.
2   Help the rest of the company improve sales of existing products.

Thus, two financial strategies are listed in Figure 3.6. As suggested in Figure 3.5, *Inventrix* R & D could also pursue a productivity enhancement strategy – improving the company's bottom line by doing things more efficiently. However, *Inventrix* decided that – given the goal of improving innovation – this could be a counterproductive strategy. Since the organization was already risk-averse, productivity improvement efforts might give employees an excuse to avoid trying new things – and risking failure. Besides, no one asked R & D to increase productivity, only to become more innovative. Again, remember the stated goal!

Having identified financial strategies, R & D management looked at how to satisfy customers. Figure 3.5 suggests three possible approaches:

1   Go for operational excellence, in which *Inventrix* seeks to become the market leader.
2   Get close to its customers and create trusted relationships and a top-tier image.
3   Create best-in-class products.

In *Inventrix*'s situation, best-in-class products and intimacy with its physician customers were selected. If R & D could develop or provide a product that was truly unique or addressed an unmet patient need, it might find itself with a blockbuster drug. On the other hand, if R & D could support marketing and sales with ammunition that would help physicians understand how existing *Inventrix* products could help their patients, perhaps sales of existing drugs could be markedly increased.

Next, R & D looked at their own organization and asked what changes in internal performance might lead to succeeding at its customer and financial strategies. Typical internal performance strategies include operations/logistics, management, innovation, regulatory and environmental (see Figure 3.5). In the case of *Inventrix* R & D, three strategies were selected: acquisition, innovation, and physician relationships. Finding new compounds to acquire would shorten the lead time to market, especially if the compound were already undergoing human trials. On the other hand, creating innovations that might enhance existing products (or could be used to create partnerships with other companies) could also lead to new products or expanded indications. Finally, since R & D has a different relationship with physicians than does the sales force, enhancing physician relationships could lead to better information transfer, more attention and possibly higher prescribing rates.

Finally, R & D management looked at the organizational issues that would need to be addressed in order to achieve success at the other three levels. As discussed in Figure 3.5, these issues typically include culture, technology, competencies and employee well-being. *Inventrix* settled on five key issues (see Figure 3.6). Let's start from the left side of Figure 3.6:

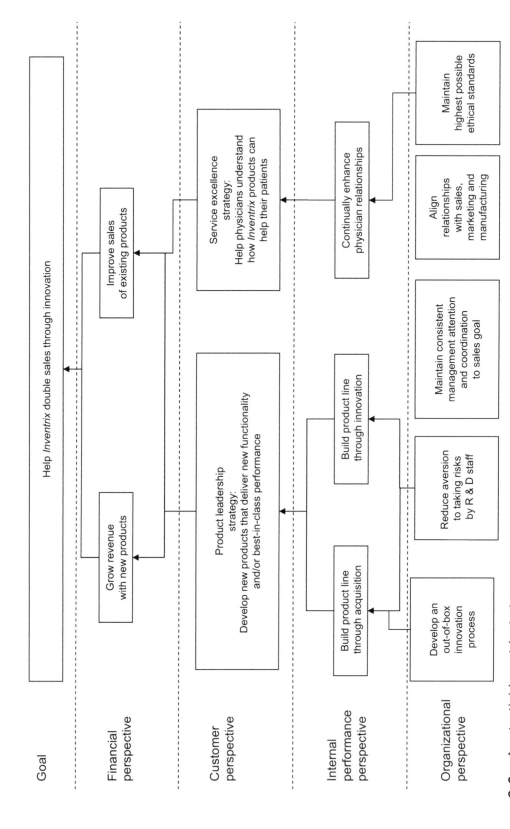

Figure 3.6 A potential *Inventrix* strategy map

- The first two issues (an out-of-box innovation process and reduction in risk aversion) focus on increasing the number and quality of innovations within R & D.

- The third issue (consistent management attention) focuses on the need to have management constantly pay attention to the goal. If management takes its eye off the ball, the rest of the organization will feel that the pressure is off and they can go back to business as usual.

- The fourth issue (relationship alignment) deals with the fact that R & D often has different priorities and goals than other parts of the company. Alignment must be maintained so that everyone stays in synch and no one is unintentionally undercutting another group.

- Finally, the last issue (high ethical standards) seeks to reinforce *Inventrix*'s history of high moral and ethical conduct and maintain it in the face of the formidable financial goal.

Note that two of the organizational growth strategies are not directly linked to any specific internal performance, customer or financial strategies. This is because these two organizational growth strategies directly impact *all* of the other strategies. In general, however, there is a direct linkage from organizational growth strategies, to internal performance strategies, to customer strategies, to financial strategies and finally to the goal. It's very important that you are able to clearly draw these arrows and linkages. If you can't then the people in the organization who have to execute the actions related to these strategies will never understand how they relate to the goal, and the line of sight depicted in Figure 3.3 will be lost.

CROmed

CROmed has the following characteristics:

- *CROmed* provides complete clinical trial services to pharmaceutical R&D organizations, but has distinguished itself because it has succeeded in cracking the enrollment problem.
- *CROmed* currently works in only one therapeutic area (TA), and now wants to expand to other TAs.
- Management is aiming for a 30 percent per year sales increase – triple the growth rate of previous years – in order to achieve $50 m in sales two years from now.
- The company is depending on its unbroken string of enrollment successes to gain customers and projects.
- However, *CROmed* must overcome major quality problems in order to grow.

A strategy map for *CROmed* might look like Figure 3.7. The *CROmed* goal is straightforward: increase sales by 30 percent per year for at least two years. *CROmed* management reasoned that there are two ways to do that: first, move into other TAs to allow room for faster growth; and second, increase profitability through increased quality

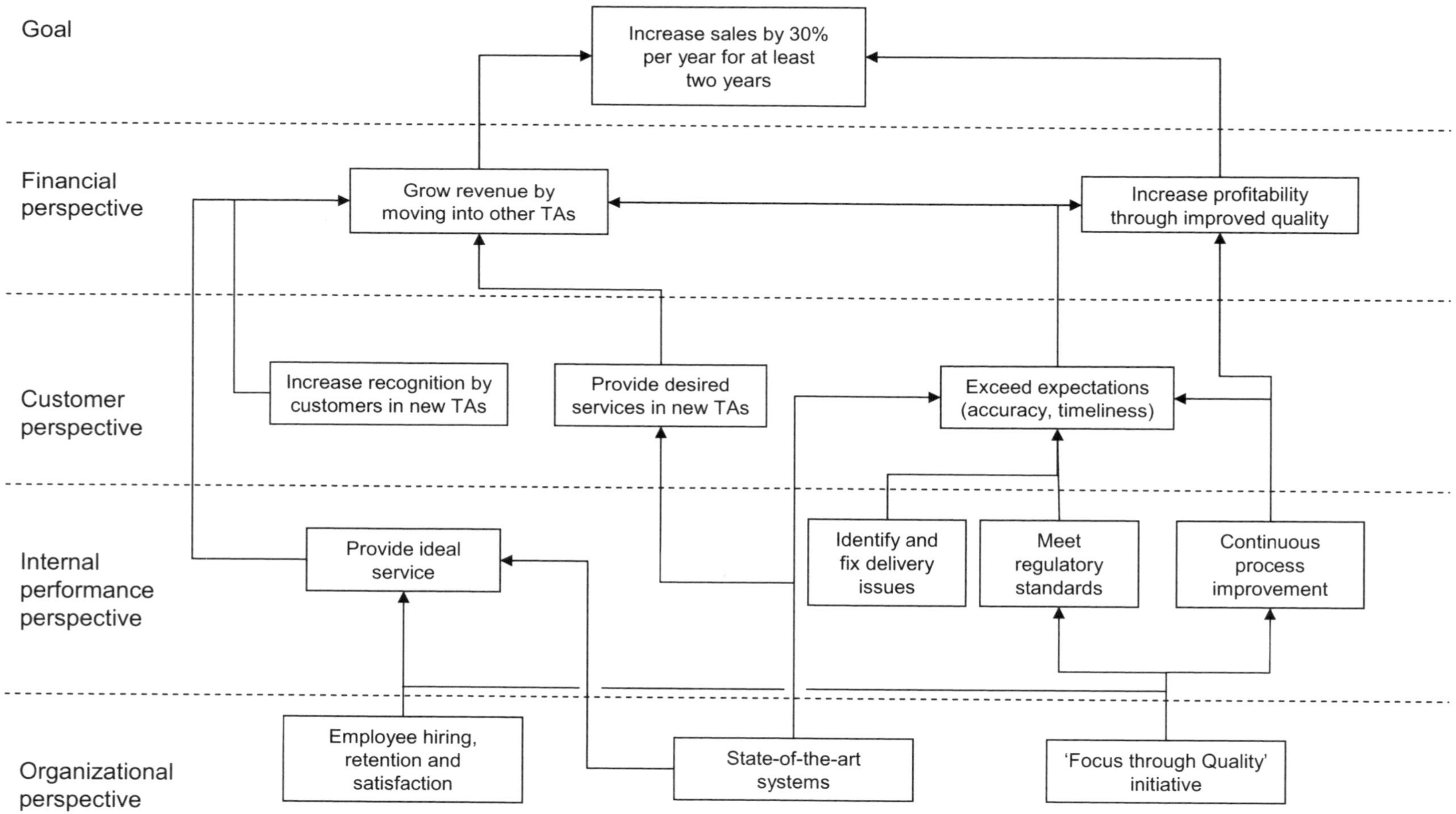

Figure 3.7 A potential *CROmed* strategy map

(*CROmed* realized that by increasing profitability, they could reduce prices and become more competitive, thereby winning more business).

In the customer perspective, *CROmed* delineated three specific strategies. To achieve growth in new TAs, *CROmed* will have to become recognized by customers in those new TAs, provide desired services (for example, by making sure that the *CROmed* contract research associates have experience in the new TAs or in providing specialized capabilities such as imaging) and exceed expectations on each project. To achieve increased profitability, *CROmed* plans to focus on exceeding expectations in all aspects of every project. This is a major shift from their traditional approach in which they focused only on enrollment.

The internal performance required to obtain these customer strategies now starts to become clear. First, *CROmed* must define and provide ideal service in each project. Second, they must meet all regulatory standards. This becomes especially important as they move into new TAs. Third, they must identify and fix all of the delivery issues that are causing them trouble in their current projects. Finally, *CROmed* must institute a system of continuous process improvement in order to increase quality.

After thinking about all of these strategies, *CROmed* management identified three major organizational growth strategies that they must execute. First, they must focus on employee hiring, retention and satisfaction. The quality problems have been taking their toll on employee morale, and there have been some key resignations. Employees must be made to feel more empowered and satisfied. Also, new hires must see quality as well as growth as a primary objective. Second, *CROmed* must bring all of its systems up to the state of the art. This will be necessary if the company is to move into new TAs and grow substantially. Finally, *CROmed* management must institute – and maintain – a focus on quality. The proposed 'Focus through Quality' initiative is meant to truly change the way *CROmed* personnel – and particularly *CROmed* management – looks at and improves quality.

This *CROmed* strategy map illustrates the importance of goal selection with regard to the identification of strategies. If you refer to the company description in Chapter 1, you'll notice that the major roadblock to *CROmed*'s growth could be defined as the quality of their product. Hence, we could have selected a goal that focused exclusively on quality. That would have driven a set of strategies that were all focused on quality as well. Instead, the goal in Figure 3.7 focuses on sales growth. As a consequence, roughly half of the strategies listed are focused on growth. It could be argued that this dual focus on quality and growth dilutes the organization's focus and divides its attention. So, if you want to focus your organization's attention and align all of its priorities, you may want to keep your goal fairly narrow to keep people from 'taking their eyes off the ball'.

BioStart

BioStart looks like this:

- 100-person biotechnology startup, with plans to grow to 150 people this year.

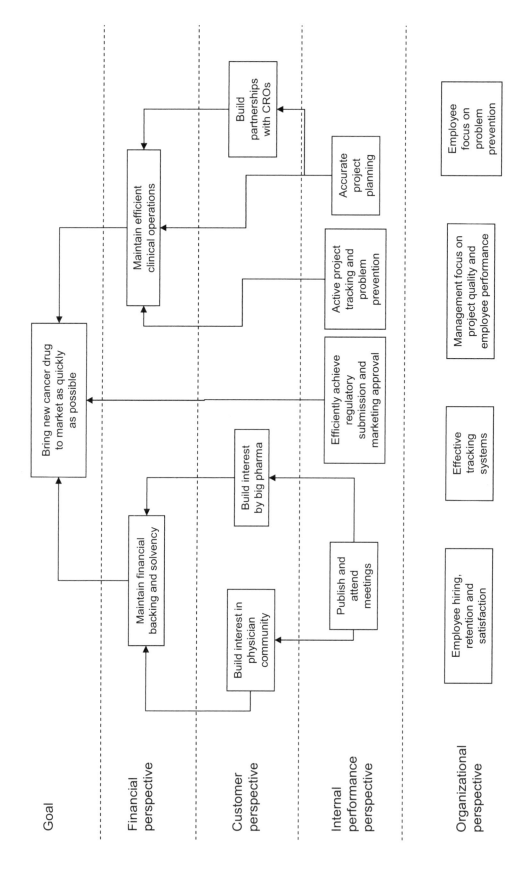

Figure 3.8 A potential *BioStart* strategy map

- Several promising new oncology NCEs that seem likely to turn into blockbusters.
- Now getting ready to go into Phase II and is thinking about a combined Phase II/III trial.
- Financed with venture capital.
- Growth in personnel and workload is taking its toll on focus and efficiency. Separation of functions means that information often gets bottled up within a functional group.
- With three to four clinical trials going on at once, it becomes essential to ensure that everything is going well and that problems are caught before they blow up.

A strategy map for *BioStart* might look like Figure 3.8. *BioStart* has a more straightforward goal than most other companies: in order to survive, it will have to bring one or more of its NCEs to market as quickly as possible. In order to do this, there are two fairly clear financial strategies: maintain financial solvency and work to make its clinical operations as efficient as possible in order to conserve cash. We could postulate a third financial strategy – to find a deep-pocketed pharma that would invest in *BioStart* or purchase rights to its products – but *BioStart* is wary of its big pharma roots and wants to remain as independent as possible for as long as possible.

In order to achieve its goal and financial strategies, *BioStart* will have to look at its two potential customers – physicians and big pharma companies – and begin courting them to assure that its drug will be well received once it is approved and that the marketing arm of a big pharma company will help sell the drug to physicians and patients. However, those two customer strategies only apply to one of the two financial strategies. As *BioStart* management looked at its strategy map, they realized that there is another 'customer' out there: CROs. While CROs are not technically a customer, *BioStart* execs realized that CROs will be an indispensable part of its clinical operations and that many CROs will be wary of committing resources to support a startup. Hence, they decided that they must court the CRO community in order to get preferential treatment for their protocols. The result is the customer strategy on the right side of Figure 3.8, which deals with establishing effective CRO partnerships.

Moving to the internal performance perspective, *BioStart* realized that they must do four things. First, they must increase their visibility in the community as much as possible. Their plan to implement this involves publishing papers and attending oncology meetings. Second, they must develop a comprehensive approach to achieving successful regulatory submissions and marketing approvals (something they haven't yet done as a startup). Third, they must plan their projects accurately and effectively. Additionally, this planning must be done in partnership with their CRO partners. And fourth, they must have an active system for project tracking and problem prevention.

These internal performance perspectives led *BioStart* management to identify in turn four key organizational perspectives:

- First, they realized that they would need effective tracking systems, especially ones that would allow everyone to track all aspects of each trial and instantly pinpoint potential problem areas.

- Second, they recognized that they would have to hire and retain people that could operate well in the new, bigger organization. The early hires had all been independent thinkers who had eschewed large, bureaucratic organizations. These people were (and are) great at thinking 'out of the box', but they also hated any kind of organizational constraints. Now, with more employees, some level of consistency and bureaucracy would clearly be necessary to ensure clean scientific results. New hires would have to have more discipline.

- Third, management realized that a larger organization would require much more management attention to, for example, project quality and employee performance; things that had previously been done on an informal basis.

- Fourth, management would have to establish a system of problem identification and prevention to ensure project success, and would have to build it into the fabric of the organization.

Note that the four organizational growth strategies underpin all of the internal performance strategies, and thus there should be arrows from these four to all of the other strategies on the diagram. For ease of reading, those arrows have been left out.

VirtuPharm

VirtuPharm has the following characteristics:

- Mid-sized, US pharma with an established set of marketed compounds and an adequate pipeline.
- R & D has always been done in-house, with ad hoc support from CROs.
- Strategic partnering with a foreign pharma has produced an expanded pipeline and fear of a Food & Drug Administration (FDA)-mandated withdrawal of their primary drug has created sudden schedule and cost pressures.
- Long-term CRO partnerships have become a sudden management priority.

A *VirtuPharm* strategy map for pharma–CRO partnerships might look like Figure 3.9. In this case, the stated goal of creating pharma–CRO partnerships represents only a piece of the company's overall goal, so this strategy map is really a subset of a larger map. The overall *VirtuPharm* goal would be something like 'growing the company despite the threat of an FDA-mandated withdrawal', and the pharma–CRO partnerships goal shown would actually be an internal operations strategy on that larger map. Hence, this example shows how strategy maps and metrics can be developed for a single piece of a larger strategy.

To achieve the pharma–CRO partnership goal, *VirtuPharm* realized that it would have to have more cost-effective operations, which would require a more efficient and cost-effective CRO. But *VirtuPharm* also realized that – if it is to have a long-term relationship with a CRO – the CRO must remain solvent and make a profit. Squeezing the CRO too tightly would create short-term savings but would risk driving it out of business. That

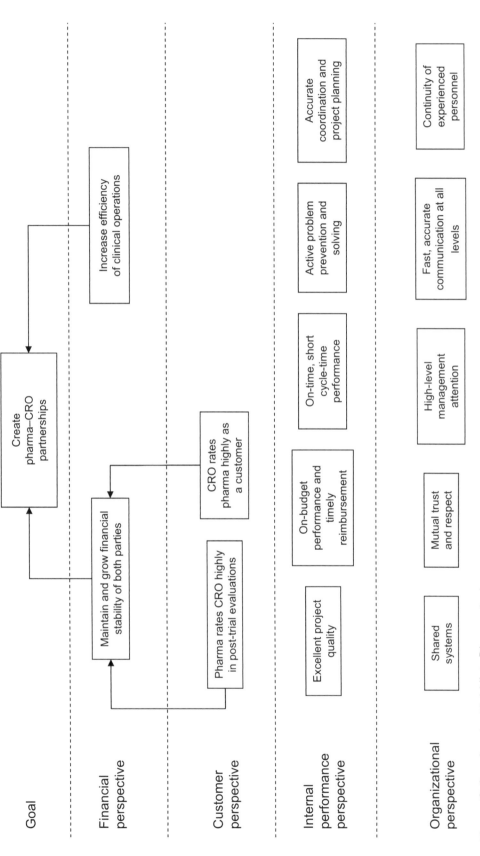

Figure 3.9 A potential *VirtuPharm* strategy map

CREATING THE
FOUNDATION
USING
STRATEGY MAPS

would compromise the 'learning curve' gains that having a long-term partner provides (see Figure 1.1) and would simply force *VirtuPharm* to find and train a new CRO. Thus, the financial strategies shown in Figure 3.9 provide for continuing financial stability and growth of both parties, in addition to increased clinical operations efficiencies.

Looking at the customer perspective, *VirtuPharm* management reasoned that not only is *VirtuPharm* the CRO's customer, but in many ways the CRO is *VirtuPharm*'s customer as well. This created two customer strategies based on post-project evaluations.

Next, management looked at the internal performance strategies required to achieve the customer and financial strategies as well as the goal. Please note that since the various strategies developed flow to multiple customer and financial strategies boxes, the myriad of arrows were left off the diagram. *VirtuPharm* focused on quality, timeliness, cycle time and efficiency/cost performance strategies, as well as processes for planning and actively finding, solving and preventing problems. Then, *VirtuPharm* management sat down with several CROs to discuss cultural and organizational issues. It quickly became clear that five organizational growth strategies would be required in order to forge successful partnerships. First, the partners would need shared systems and continuity of experienced personnel in order to achieve success. But more fundamentally, each organization would have to maintain continuity in their work forces, each must learn to trust and respect the other, management would have to pay regular attention to the relationships, and constant communication would have to be achieved at all levels.

SUMMARY

We've now delved into the relationship between goals, strategies, actions and metrics. In Figure 3.2, we described that relationship. We also talked about creating line-of-sight strategies that allow individuals 'in the trenches' to relate to broad, corporate-level goals. Then we discussed how to build a strategy map that links financial, customer, internal performance and organizational growth strategies to your goal.

The four examples above show different ways that strategy maps can be applied to organizational problems. I recommend that you now take some time to build a strategy map for your own organization. Use the blank strategy map in Figure 3.10, or draw your own.

● First, state a single, overriding goal that your organization must achieve. Make sure that the goal is achievable, but not easily so. If it's too easy to achieve, then the problem will be solved before you can even develop the metrics. If it's impossible, you'll just frustrate the entire organization.

● Second, list out several financial strategies that will be required to achieve the goal. Increasing revenue, increasing profitability, reducing cost or increasing efficiency are all good financial goals.

● Third, identify who your customers are in the context of the goal, and determine the customer strategies required to make your financial strategies a success. In the world of R & D, it is often difficult to identify your customers, but you do indeed have them and it's important to figure out which customers truly have an influence on whether you achieve your goal.

● Fourth, figure out what internal performance strategies will help you achieve your customer and financial goals. Most people find it easy to list lots of internal performance strategies, so the trick here is to work out which ones have the most impact on your goal.

● Fifth, look critically at your organization and decide what improvements will be most helpful in achieving the various strategies and goals you've already listed.

● Finally, draw arrows from the various strategies to their outcomes to give yourself a clear map of how one strategy affects another.

Goal

- -

Financial
perspective

- -

Customer
perspective

- -

Internal
performance
perspective

- -

Organizational
perspective

Figure 3.10   A potential strategy map for your organization

CHAPTER 4

Developing Performance Metrics

Some years ago, the CEO of a manufacturing company was trying to figure out how to increase profitability. His company assembled products from parts provided by suppliers. When he looked at the cost of production, it was clear that two thirds of his cost was in purchased parts, with the remaining third consumed by his own assembly process. It didn't take him long to realize that he could make much greater improvements in profitability by reducing procurement costs than by reducing assembly costs. After all, cutting assembly costs in half would decrease his overall production cost by 16.5 percent, while cutting his procurement costs in half would reduce overall production costs by 33.3 percent! So he went to his procurement director and asked him to start measuring procurement costs on a part-by-part basis, and set a goal of halving the company's procurement costs. Then he sat back – with a satisfied grin – and awaited the superlative outcome he knew would occur.

Sure enough, the procurement numbers began to trend downward (he had incentivized his procurement director with some serious bonus money, so he was delighted). But oddly, his overall production costs remained the same – and then began to climb! Pretty soon, the ratio of procurement to assembly costs had changed from 2:1 to 1:1. What was going on?

He went down to the manufacturing floor to try to get to the bottom of the mystery and almost tripped over a large pile of scrapped parts. After some investigation, he figured out the problem: In order to reduce the cost of purchased parts, the procurement director had relaxed part quality requirements and delivery schedules. Parts started coming in late, which slowed down the assembly process and reduced the number of products assembled per labor hour. The parts that did arrive were often of lower quality. Many of them had to be scrapped outright or had to be reworked along the assembly line to make them fit.

What happened here? The CEO realized that he had 'suboptimized' the production process. He had focused on only one part of the process (procurement) rather than on the overall process (procurement + assembly). The result was that one part of the process had benefited to the detriment of the rest. In addition, he had only measured one aspect of procurement (cost) and had neglected to measure other aspects (delivery cycle times and part quality), thereby creating an 'unbalanced' view of the situation. The procurement director had taken advantage of the lack of balance and had traded cost (which was being measured) for quality and cycle time (which weren't being measured). Wisely, the CEO did not blame the procurement director. It was obvious that the director was simply responding to the rules that the CEO had set up and was working in his own best interest.

You may be saying to yourself: 'Why are manufacturing examples being discussed? Isn't this book about R & D?' Indeed, many researchers believe that R & D and manufacturing have little in common; but while pharmaceutical R & D doesn't churn out toasters or cars, it does produce products (NCEs, devices, clinical data, regulatory filings) and it does have repetitive operations (starting clinical trials, enrolling patients, and so on). Thus, like the example above, it is possible to suboptimize the R & D process and create unbalanced views of what's going on. Here are a couple of examples:

- There has been a spate of articles recently about the rise of combinatorial chemistry and high-throughput screening in the drug discovery process. Many companies invested heavily in this equipment in the hope that the vast number of molecules the equipment could create and test would reveal new blockbusters at a blistering pace. However, the results have been discouraging. Why? Well, some observers point out that the equipment blindly tests oodles of molecules that a knowledgeable researcher would likely reject out of hand (too large, insoluble in water, not manufacturable), while skipping less obvious combinations that a human expert might try 'on a hunch'. In effect, the equipment shifted the discovery process from an emphasis on quality to an emphasis on volume – not unlike the tradeoff that the manufacturing CEO made in focusing on procurement costs – and the result has been suboptimization.

- When development groups begin a protocol, they work hard to develop the best protocol possible. However, they rarely have enough time to consider all the issues, so they focus on the science and frequently leave it to the operations group to make the protocol work out in the field. Of course, by the time the operations group gets the protocol, both the company and regulatory agencies have already approved it, so making a change is both expensive and time-consuming. More often than not, the development group has severely restricted the patient population to create a homogenous group, or added difficult tests and procedures in order to eke out a little more data. The operations group then has to contend with these difficulties, either by expending tremendous effort to recruit and retain sites and patients, or by amending the protocol in order to obtain a satisfactory sample size. In this case, the company has suboptimized by emphasizing protocol development cycle time over protocol quality and has sacrificed overall clinical trial cycle time. It's common to see R & D measures of protocol development cycle time or on-time protocol release, but it's almost unheard of to have a measure of protocol quality. Yet, recent work by my consulting company has revealed that perhaps two thirds or more of protocol amendments that accrue in a typical protocol could actually have been foreseen had the development group taken the time to carefully simulate protocol execution and optimize the protocol.

In both of these examples, organizations have chosen to take what appears to be the easier path and in the process have sacrificed overall performance. They have also chosen to measure things that are easy to measure (cycle times, on-time performance) and have ignored more difficult metrics (quality). The result has been to skew organizational behavior in favor of what *is* being measured (get more compounds screened or get the protocol written faster) as opposed to what *should be* measured (high-quality protocol execution or high-probability molecules).

So, how do we determine the *right* metrics to use? In Chapter 2, we discussed setting appropriate improvement goals, and in Chapter 3 we developed strategy maps that helped us identify the major issues that the organization should address. Those goals and maps will now be our guide to developing the right set of metrics. We'll attempt to identify one or more metrics that will measure our success in implementing each of the individual strategies. Because each of those strategies has been linked to our overall goal, the set of metrics we create will in effect be measuring progress toward our goal; exactly what we had intended.

Now, it's certainly possible to skip the strategy mapping effort and try to develop metrics directly from our goal. It certainly seems like an easier path! In fact, it may seem that the entire goal setting and strategy mapping effort is a waste of valuable time. Why not just get right to the business of building a metrics set and collecting data. The reality is that there are hundreds, or even thousands, of potential things to measure, and only a few of them are ideal for measuring progress toward your goal. Without a disciplined process for selecting metrics, I can guarantee that you'll end up with a pot full of the wrong metrics. The odds of picking the right metrics from such a large number of possibilities is very small indeed. You might think that you can avoid this problem by getting input from various people around the organization. That's a good idea, but like the manufacturing CEO and procurement management, the odds are that you'll end up with metrics that will lead to suboptimization. So a word of caution here: Don't skip the goal setting and strategy mapping steps in an effort to 'cut to the chase'. In trying to hurry through the process, you'll end up costing yourself both time and money.

When we developed our strategy maps, we started with financial strategies and then successively developed customer, internal performance and organizational growth strategies. We could do the same with our metrics: start at the top – with the financial perspective – and work our way down, creating metrics for each strategy as we go. However, as scientists and R & D professionals, we tend to be most comfortable dealing with day-to-day operations and quantitative data – in essence, with data that we can clearly gather and affect. So it is often easier to begin with the internal performance perspective than with any of the other three. Financial aspects are quantitative, but they measure results that can be months or years downstream from the work being performed, so it's difficult to know how our day-to-day work affects them. Customer measures often seem subjective and unrelated to our daily work; we can work hard and do a great job, only to have a dissatisfied customer as a result of a communications failure or an unanticipated glitch. Organizational growth measures tend to be even more subjective and may seem to have nothing at all to do with the work we're engaged in. All four aspects are critical to long-term success, but the internal performance metrics are the easiest to grasp, so we'll start there.

DEVELOPING
PERFORMANCE
METRICS

We'll spend this chapter dealing exclusively with internal performance metrics, and then treat the other three perspectives in Chapter 5.

GOALS OF PERFORMANCE METRICS

In the US, American football teams measure everything they do. They have top-level measures, such as how many points they've scored and how many points have been scored against them. And then they have detail measures, including how many yards they've gained on each down, how many yards they've gained by running vs. passing, how effective each offensive and defensive play has been, and so on. Virtually everything about football is measurable. Even the field itself is marked off in increments to facilitate measurement of forward (or backward) progress. Between the field, the scoreboard and the commentators, it's easy to figure out at any given moment how well or poorly the team is doing and how far the team has to go to obtain victory.

As you begin to examine your performance, you'll find yourself asking questions that are similar to those of a football team. The answers are definitive and numerical:

- Just how good (or bad) are things?
- How much better can our performance get?
- Are things improving or getting worse?
- What will it cost us to make various improvements?

In fact, there are many 'commentators', 'coaches' and owners who are watching and measuring the performance of your R & D organization: Financial markets such as New York's Wall Street or London's Square Mile use financial data to evaluate the viability and potential of your company and determine whether performance is heading in the right direction. Meanwhile, your organization's senior executives have their own measures which track performance and allow them to make decisions based on trends, pipeline data and so on. As you develop your performance metrics system, you'll have to develop your own measures that will allow you to determine what's happening and whether you're headed in the right direction.

That's what a performance metrics system is for: it's a set of measures designed to answer five questions:

1 How is the system performing now?
2 Which aspects of your system are performing poorly and need improvement?
3 Which aspects of your system are just fine?
4 What is ideal performance?
5 Are your improvements having the desired effect?

You can gauge the success of your performance metrics system by whether it truly addresses each of these five questions. By systematically building metrics based on goals and strategies, you should have no trouble addressing all five of these questions.

TYPES OF PERFORMANCE METRICS

Earlier in the chapter, we discussed the issue of balance in a metrics system. In particular, we mentioned that suboptimization (for example, focusing on cycle time or cost while ignoring quality) can lead to disastrous results. For this reason, it's important to recognize four different types of metrics, and be able to readily distinguish them:

- Timeliness – a measure of success in meeting a deadline or customer requirement
- Cycle time – a measure of the time from customer request to delivery

---

Timeliness (T)

- On-time protocol completion
- On-time enrollment
- On-time report completion

Quality (Q)

- Number of blank data fields per CRF page
- Number of queries per data point
- Number of report errors per page

Cycle time (CT)

- Average time to initiate a site
- Average time to enroll 10 patients
- Average time to close a database

Efficiency/cost (E)

- Cost per enrolled patient
- CRFs harvested per CRA-day
- Cost per clean CRF page

Figure 4.1 Metrics examples

- Efficiency/cost – a measure of the output per unit of resources used (labor, dollars and so on) or its inverse (resources required per unit produced)
- Quality – a measure of the number of errors or defects relative to a customer's requirements.

Figure 4.1 lists some example metrics for each type.

It's important to realize that not all metrics fall into these four categories. Try the quiz in Figure 4.2 to make sure you understand which metrics fall into which categories – and which metrics fall into none of them. The answers to the quiz can be found in Figure 4.3.

The metrics in Figure 4.2 are typical of those found in pharmaceutical R & D organizations. What's amazing is how many of them don't fit in any of our standard categories! Why is this important? Because a metric such as 'number of sites initiated' really tells us nothing. It doesn't pass the 'so what' test ('Now that we know the metric value, so what?'). In the case of the number of sites initiated, we don't know if the value – say 30 – is a good number or bad number. If it were 60, would that be better? Maybe, but only if we needed more than 30 sites! On the other hand, if we convert the metric to 'percent of sites initiated on-time', we immediately understand what's going on: anything less than 100 percent is undesirable, while a number such as 30 percent would prompt immediate action. Similarly, 'number of patients enrolled' – although perhaps the most common of all clinical operations metrics – tells us virtually nothing. Changing the

Fill in the category for each metric below. Fill in not applicable (N/A) if the metric doesn't fit into any of the four categories. If the metric is 'N/A', try redefining the metric so it does fit in one of the four categories.

Metric	Category (T, CT, Q, E or N/A)	If 'N/A', redefine the metric so that it does fit in a category
Number of sites initiated		
First site initiated on-time		
Number of patients enrolled		
First patient enrolled on-time		
Number of CRFs harvested		
Number of queries generated		
Number of queries outstanding		
Time from last patient, last visit to report		
Key: T timeliness; CT cycle time; E efficiency; Q quality		

Figure 4.2   Metrics categorization quiz

metric to a timeliness metric (percent of patients enrolled on-time) or to an efficiency metric (number of patients enrolled per site-day, where a site-day is our unit of effort) gives us a great deal of information to work with and immediately tells us whether remedial action is required. Even more revealing is 'actual enrollment per site-day vs. forecast'. Such a metric would tell us not only how we're doing from an efficiency viewpoint, but also whether we are on schedule. It's really a combined efficiency/ timeliness metric, and could be classified under either type. Figure 4.3 lists some possible alternative definitions for those metrics that don't fit into any of our four categories.

You may be wondering at this point why I've defined four categories of metrics, instead of the traditional three: cost, quality and time. Here's my reasoning: timeliness and cycle time do in fact use exactly the same data to develop their metrics, as cycle time is simply the elapsed time between two milestones. However, in R & D there are many processes where *how fast* something is done is more important than *when* it is done, while in other cases the reverse is true: it's more important to be done on-time than to be done quickly. To clarify, it is important to remember that the execution of a clinical trial is a fairly linear process, for example you can't execute the visit schedule until you have initiated the site and enrolled the patient. So if site initiation is late, it's likely that all of your subsequent milestones and their associated timeliness metrics will also be late. The first late timeliness metric gives you some important data, but the subsequent late metrics give you no additional data; they're all late because the first one was late. On the other hand, the cycle time for each subsequent step is additional, useful data. Even if the site initiation was late, it's still valuable to know how quickly you're enrolling patients, completing CRFs, and so

Metric	Category	If 'N/A', possible redefinition
Number of sites initiated	N/A	Percent of sites initiated on-time (T) Average days to initiate a site (CT)
First site initiated on-time	T	
Number of patients enrolled	N/A	Percent of patients enrolled on-time (T) Actual enrollment per site-day (E) Actual enrollment per site-day vs. forecast (E or T)
First patient enrolled on-time	T	
Number of CRFs harvested	N/A	Number of CRFs harvested per CRA-day (E) Average time from CRF completion to arrival at data management (CT)
Number of queries generated	N/A	Number of queries per CRF page (Q)
Number of queries outstanding	N/A	Average time to close a query (CT)
Time from last patient, last visit to report	CT	
Cost per clean CRF	E	
Key: T timeliness; CT cycle time; E efficiency; Q quality		

Figure 4.3   Answers to the metrics categorization quiz

on. As another example, closing and locking a database is something that must be done quickly, no matter whether the trial is on schedule, ahead of schedule or behind schedule. Thus, in this instance, cycle time is critical and timeliness is relatively unimportant. In general, I have found that milestones at the beginning and the end of a process (for example, on-time protocol release, on-time CRO contract signing, on-time final report) are usually best measured with timeliness metrics, because it is critical that these deliverables not be late. On the other hand, intermediate milestones (for example, site initiation, enrollment, data collection, database close/lock) are usually best measured with cycle time metrics because the exact dates when they are completed are less important than how quickly the tasks were accomplished.

So the first principle of performance metrics selection is this:

Principle 1: Select metrics that can be categorized as timeliness, cycle time, quality or efficiency.

Why not measure both metrics that fall into these four categories and other metrics as well, you ask? The simple answer is this: Of course you can measure as many things as you wish. However, having too many metrics can become quite confusing. Better to have a limited number of critical metrics that everyone can remember and focus on, than a huge number of metrics that capture all the data but obfuscate the critical issues. And that leads us to the next issue...

CHOOSING THE RIGHT METRICS

Here's an old joke:

> A man comes out of a restaurant late one evening and finds his friend wandering
> around under a streetlamp peering intently at the ground.
> 'What are you doing?' he asks.
> 'I lost my keys while unlocking my car,' is the reply.
> 'But I saw your car parked half a block away.'
> 'I know. But it's dark over there. At least over here I can see!'

Most metrics systems are constructed like this joke: they measure what's easy to measure
rather than what's important to measure. For instance, it's easy to measure timeliness – and
therefore cycle time measures – because we can just compare the actual date to the date
we planned to be done (forecast date). But if the problem was one of quality or resources,
knowing we were late doesn't really tell us much about the cause; it only verifies what we
probably already knew. So the second principle of performance metrics selection is:

Principle 2: Measure what's important, not what's easy.

A corollary to principle 2 is 'measure causes, not just effects'. In clinical trials, most
organizations will measure the data collection and trial execution steps. This data is easy to
collect, but it misses the fact that problems in trial execution and data collection are often
the result of inadequate planning, hurried site selection and overly complex protocols.
However, those aspects of the protocol are rarely measured. Why? Because it's easy to
measure the execution parts of the clinical trial (site performance, data collection, database
closure), but it's hard to measure the planning and preparation parts of the trial (protocol
development, site selection). We tend to measure this way even though we know that
problems in clinical trial planning and preparation are often the cause and problems in
execution are the effect. By the time we measure the effect, our only recourse is damage
repair, which involves expending more resources to try to meet already-compromised
milestone metrics. If we could measure quality at the causal stage, we could eliminate the
potential problems rather than having to take remedial action. Our inability to measure
adequately the quality during the planning and preparation of the trial causes us to
suboptimize: we reduce cycle times during the planning and preparation at the expense of
quality, cycle time, efficiency and timeliness during execution. Thus:

Principle 2a: Measure causes, not just effects.

Next, it's important to keep a balanced set of performance metrics. Why? Think about it
this way: If we measure and improve only timeliness, we would be ensuring that we meet
our contracted deadlines. But we can always achieve this if we just start far enough in
advance. Give me a few years (or decades) and I can meet virtually any milestone! In
essence, timeliness can always be achieved at the expense of cycle time. This is
demonstrated in Figure 4.4, where pushing on one set of metrics unbalances the balloon
circle, and it deforms on the opposite side.

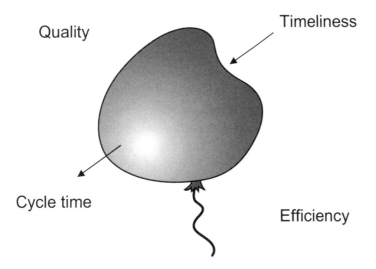

Figure 4.4 Unbalanced metrics distort the balloon

Now, we could measure and improve *both* timeliness and cycle time; get things done on time and do it fast. We could do this by throwing additional resources at the problem, thereby reducing efficiency. With infinite resources we can do almost anything both on time and as quickly as you want. So now we're achieving timeliness and cycle time at the expense of efficiency.

Finally, we can measure and improve all three values – timeliness, cycle time and efficiency – if we allow ourselves to produce junk: poor quality. So we need to measure quality as well.

Thus if you think of your system as a balloon, pushing on one side simply allows the balloon to expand on the other side. The only way to improve overall performance is to push on the balloon from all directions, thereby decreasing its size and improving overall performance, as in Figure 4.5.

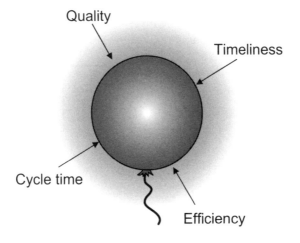

Figure 4.5 Balanced metrics produce symmetric improvement

So the third principle of performance metrics selection is:

Principle 3: Create a balanced set of metrics.

A corollary to principle 3 is the law of unintended consequences, or 'what gets measured gets fixed'. As we just described, when you measure timeliness and tell your organization that's what's important, your employees will do whatever they can to improve timeliness. It's really true that when an organizational measure is initiated, the organization will do whatever it takes to improve that measurement. However, most people will try to take the easiest route to improving that measure. In the case of improving a timeliness measure, the easiest thing they can do is start earlier or put more people on the job. Thus, concentrating on milestones will cause timelines to expand and costs to go up. A good example of this is the airline industry in the US. A few years ago, the Federal Aviation Administration (FAA) started publishing the on-time arrival of all flights – any flight arriving within 15 minutes of its scheduled arrival was deemed on time – in reaction to myriad passenger complaints about late flights. The industry immediately reacted and US airline flights are now routinely on time 90+ percent of the time. How did they accomplish that? They stretched out the flight times! Suddenly, a St Louis to Chicago flight increased from 50 minutes to an hour and 5 minutes – with longer times during peak hours. St Louis and Chicago didn't move 30 percent farther apart, the airlines just gave themselves 30 percent more cycle time leeway to meet the timeliness target. You might say that at least flight arrivals are now more predictable – and in fact passenger complaints to the FAA about late arrivals have decreased dramatically and that relieves passenger stress – but the reality is that no fundamental improvement occurred at the airlines. Flights still arrive later than they need to; we're all simply acknowledging that fact rather than fixing it. On the other hand, if the airlines were not just measured on on-time arrivals, but also flight times (cycle times) and service quality, they would have to make fundamental changes in their route structures and service models to meet the combined set of metrics.

Principle 3a: Measure not only what you want to fix, but also what you don't want to get worse.

Having said all this, we know that measuring timeliness and cycle time is fairly easy, while measuring quality and efficiency is much more difficult. How are we supposed to measure the quality of a protocol or of a site selection effort? A concept called 'next operation as customer' (NOAC) can be helpful here. Figure 4.6 shows a simple protocol development process. It includes eight tasks, including two review tasks (these are ineffective, but

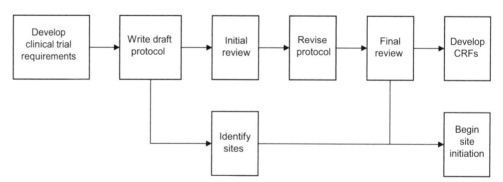

Figure 4.6   A simple protocol development process

typical, steps in most companies), as well as two site selection tasks that proceed in parallel with the protocol development. If we believe that these early tasks are the seeds of later execution problems, we need to somehow measure the quality of this process. Figure 4.7 shows a way to do this through NOAC.

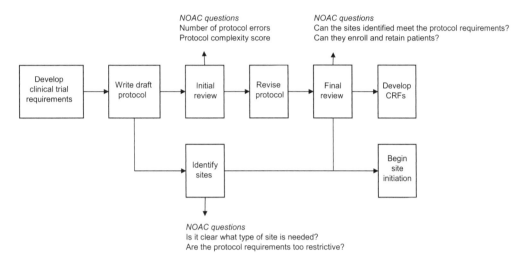

Figure 4.7 Using next operation as customer (NOAC) to understand execution problems

In essence, the NOAC approach uses a map of the process under review and asks a simple question of each of the tasks internal to the process: 'How well did the previous task perform its job?' Each task can be viewed as the 'customer' for the output of the previous task, and can therefore rate the quality of the previous task's output. In Figure 4.7, we can see that the initial review task can look at the quality of the draft protocol in terms of errors and complexity rating, while the identify sites task can ask whether the protocol requirements can be addressed successfully by the sites. These questions are never explicitly asked in most organizations. The result is that protocols are difficult to enroll or overly complex, creating enrollment problems and subsequent protocol amendments. If these metrics are captured and analyzed, protocol quality will increase and enrollment and execution cycle times will decrease. So now we have a fourth principle of performance metrics selection:

Principle 4: Use process mapping to find internal performance metrics.

A corollary to principle 4 is that internal performance metrics (those that are at the beginning or in the middle of a process) are generally predictive, whereas end-of-process metrics (those at the conclusion of a process) are generally retrospective with respect to that process. For this reason, cost metrics – which are generally end-of-process efficiency metrics – are retrospective rather than predictive metrics. They tell you what's happened in the past rather than what might happen in the future. While it's important to have retrospective metrics so that you'll know what to do or not do next time, it's also valuable to have predictive metrics that allow you to make mid-course corrections and hopefully bring your end-of-process metrics in line with forecasts.

Principle 4a: Use internal, predictive metrics when possible.

HOW MANY METRICS?

As to how many metrics, I differ from many people on this issue. Most experts feel that you create as many metrics as you need, which usually means 'use a lot of metrics'. My view is that you create only as many metrics as you actually can use – and can easily remember. I've worked with organizations that have literally hundreds of metrics. When you ask people in the organization how many they actually use, they'll thumb through the list and point to three or four that they keep tabs on. The rest they ignore. Why go to the trouble to collect, analyze, maintain and report on all that data if you use only a small fraction of it? Psychologists tell us that short-term memory holds only about seven items. Personally, I can't keep more than three or four things in my head at any one time, and I have a hard time remembering more than about a dozen. So my rule of thumb is that a performance metrics scorecard should have roughly a dozen metrics (three per axis) and no more than 18. Keeping the number of metrics to a minimum has another advantage: when you have a large number, you get trapped by the fact that *when everything is important, nothing is important*. With so many measures, it's hard for people in the organization to prioritize and keep focus. By boiling it down to a small number, it forces you to decide what's really important, and makes it clear to the organization what management truly values. So the fifth principle of performance metrics selection is:

Principle 5: Restrict your metrics to the critical few.

DEFINING YOUR METRICS

OK. So now you have a set of metrics that seem to conform to the five principles we've listed. You're done, right? Not so fast! Once you get a set of metrics that will help you in your improvement efforts, you need to define each metric carefully. Seems simple? In fact, this is one of the most difficult steps in setting up a metrics system.

Here's an example from a different industry that clearly illustrates the issue. An aerospace company wanted to measure the quality of parts they received from suppliers. This company had several geographic locations and wanted to create a consistent received parts quality metric. They decided to define their metric as:

number of 'rejection tags' written by the incoming inspection department per month.

This seemed to be a reasonable metric since inspectors check all incoming parts against the specifications, and write a 'rejection tag' any time they find a discrepancy. And the definition seemed straightforward – until they started to look at the data. It turned out that:

- some locations inspected every piece, while others inspected a sample from the incoming order
- some locations wrote an inspection tag for every discrepant part they found, while others wrote a single tag for an entire incoming order, even if they found ten bad parts in the order

- some locations wrote a tag as soon as they encountered a discrepancy, while others gave the supplier several chances to fix the parts first.

So the number of rejection tags could vary greatly depending on the rules an individual location used to write tags.

There were many other definitional problems as well. The improvement team had to go back and refine the metric definition so that all locations collected and reported consistent data.

Then there was the problem of volume. Different locations processed different numbers of parts. So a metric that showed 10 rejection tags per month meant a big problem at a location that processed 100 parts per month, whereas it was a trivial issue at another location that processed 10 000 parts per month. So the team had to 'normalize' the metric to produce consistent results:

$$\frac{\textit{number of 'rejection tags' written by the incoming inspection department per month}}{\textit{number of received parts per month}}$$

As you can see in the above example, metrics almost always need to be normalized. Normalization is the process of equalizing metrics across different times, locations, and so on. Normalization involves placing one or more factors in the denominator of the metric. In the above example, the normalization factor is the number of parts received.

If the above metric were simply defined as the number of rejection tags written, the data would have been useless. Receiving sites that had only a few parts delivered per month would have only a few rejection tags, while sites with thousands of parts delivered per month would have a great many rejection tags. Does that mean that the large site has poorer supplier quality than the small site? Of course not. By dividing the number of tags by the number of parts received, the metric is normalized and data from different sites can be compared.

The problem is the same in pharmaceutical R & D. If you're trying to measure the cycle time from last patient last visit to database lock, you'll immediately run into the problem of what is meant by database lock. Some groups will define that as the point where all cleanup is complete, but they still allow the database to be 'unlocked' if further problems are found. Others define it as the time after which the database is never changed. And there are numerous other definitions. Meanwhile, there's a volume issue here as well: you can't equate the time required to lock a 10 000-patient database with 200 data points per patient (2 million data points total) with one having only 100 patients and 50 data points per patient (5000 data points total). So you need to normalize the metric in order to be able to compare realistically across trials (compare 'apples to apples').

That gives us our last two principles of performance metrics:

Principle 6: Carefully define your metrics.

Principle 6a: Normalize your metrics.

Principle 1:	Select metrics that can be categorized as timeliness, cycle time, quality or efficiency
Principle 2:	Measure what's important, not what's easy
	Principle 2a: Measure causes, not just effects
Principle 3:	Create a balanced set of metrics
	Principle 3a: Measure not only what you want to fix, but what you don't want to get worse
Principle 4:	Use process mapping to find internal performance metrics
	Principle 4a: Use internal, predictive metrics when possible
Principle 5:	Restrict your metrics to the critical few
Principle 6:	Carefully define your metrics
	Principle 6a: Normalize your metrics

Figure 4.8 Principles of performance metrics selection

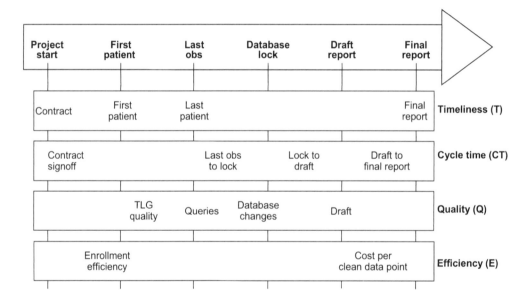

Figure 4.9 Example of a balanced set of performance metrics

Figure 4.9 shows an example performance metrics set that was developed for Searle some years ago.[1] The system obeys all six principles, although in some cases the metrics are not normalized as well as they could be.

METRICS – USUAL AND UNUSUAL

Over the years, I've compiled lists of metrics that are used by companies performing

1. 'Measuring the Work of Contract Research Organizations,' D.S. Zuckerman, C.E. Piper and M.P. Cole, *Scrip Magazine*, July/August 1998, pp. 29–30.

pharmaceutical R & D. I've run industry-wide surveys and worked with many different clients. A sampling of some of the typical and some of the more unusual metrics I've come across is shown in Figures 4.10 and 4.11.

Timeliness
    Contract signed on-time
    Protocol issues on-time
    FPFV on-time
    Enrollment complete on-time
    All data in-house on-time
    Final report on-time

Cycle time
    Protocol draft to final
    Protocol final to FPFV
    Enrollment cycle time
    LPLV to last data in-house
    LPLV to database lock
    Database lock to draft report

Quality
    Queries per CRF page
    Queries per 1000 data points
    Total queries vs. Type II queries
    Database errors after close
    Database errors after lock

Efficiency
    Cost per trial
    CRO cost per trial
    NDAs created per year
    Protocols completed per year

Figure 4.10   Typical metrics in pharmaceutical R & D

Quality
    Site quality audit problems
    Site-generated errors per CRF page
    Queries per page as a function of time (learning curve)
    CRO contract change orders (with or without scope change)
    Protocol revisions
    Protocol quality score
    Monitoring hours per CRF page
    Site quality score
    Percentage of queries which are common across sites
    Database errors found by statistics group (NOAC)
    Statistics errors found by medical writing group (NOAC)
    Missed AEs and SAEs

Efficiency
    Actual enrollment vs. forecast as a function of time
    CRFs harvested per CRA-day
    Earned value (actual vs. forecast)
    Hours to develop/debug a new database
    Total project hours per milestone

Figure 4.11   Unusual metrics in pharmaceutical R & D

CASE STUDIES: METRICS FOR EXAMPLE COMPANIES

Figure 4.12 shows the internal performance strategies that were developed in Chapter 3 for each of our four example companies.

In the sections below, we provide some example metrics for the various situations of our four example companies. Obviously, there are many more possible metrics and

DEVELOPING
PERFORMANCE
METRICS

BioStart
- Publish and attend meetings
- Efficiently achieve regulatory submission and marketing approval
- Active project tracking and problem prevention
- Accurate project planning

CROmed
- Provide ideal service
- Identify and fix delivery issues
- Meet regulatory standards
- Continuous process improvement

VirtuPharm
- Excellent project quality
- On-budget performance and timely reimbursement
- On-time, short cycle time performance
- Active problem prevention and solving
- Accurate coordination and project planning

Inventrix
- Build product line through acquisition
- Build product line through innovation
- Continually enhance physician relationships

Figure 4.12 Internal performance strategies that were developed in Chapter 3 for each of our four example companies

combinations of metrics that could be developed, depending on the circumstances, but these examples should give you a place to start when you are designing your own performance metrics.

BioStart

BioStart has the simplest set of internal performance strategies. It aims to keep its projects on track and make sure it is staying in the community spotlight. Here are some possible metrics that *BioStart* could use. The type of metric (T, CT, Q, E) is listed in bold for each.

1 Accurate project planning
   a. Percentage of projects that have a complete project plan – **Q**
   b. Number of project forecast revisions in the last three months (divided by number of active projects) – **Q**

2 Active project tracking and problem prevention
   a. Percentage of milestones in the last three months that are less than five days late to project forecast – **T**
   b. Number of problems identified per project in the last quarter – **Q**
   c. Number of queries generated per project data point for projects that have achieved database lock in the last three months – **Q**
   d. Average resources per project milestone achieved in the last three months compared to forecast – **E**

3 Efficiently achieve regulatory submission and marketing approval
   a. Time from completion of final report to regulatory submission – **CT**
   b. On-time annual reports – **T**
   c. Average personnel hours required for a regulatory submission for submissions made in the last year – **E**

4 Publish and attend meetings
   a. Percentage of 'high visibility' oncology meetings during the last three months that had a *BioStart* employee in attendance – **T**
   b. Number of papers submitted to refered journals in the last three months

We've identified 11 metrics for *BioStart*, four of which are quality, three timeliness, two efficiency, one cycle time and one that doesn't fit a category (yes, there are times when a simple count such as metric 4b is appropriate). There are certainly other metrics that could be considered, so this list shouldn't be thought of as either exclusive or exhaustive. The exact list for your organization would clearly change depending on circumstances and goals.

Note that these metrics all use a three-month 'moving average'. A moving average smoothes out variations in the data by averaging over an extended period of time – in this case the last three months. Figure 4.13 shows the impact of moving averages. Counting every data point individually (curve A) allows you immediately to see trends and changes in the data, but it introduces lots of variation, or 'noise'. Using a three-period moving average (curve B) reduces the noise so that it's easier to see trends. Using a 12-period moving average (curve D) provides even more smoothing, but trend performance tends to lag significantly behind actual changes. Choosing the right length for a moving average involves trading off the improved noise reduction against the increasing lag in trending.

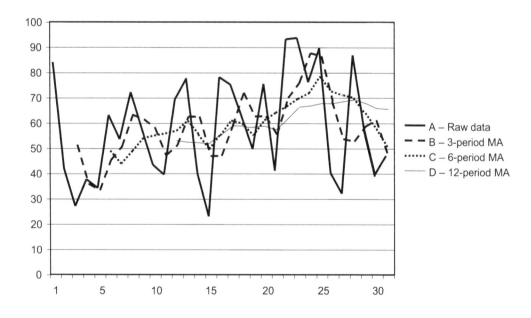

Figure 4.13   The impact of moving averages (MAs)

CROmed

CROmed has a fairly straightforward set of performance goals, although a bit more complex than *BioStart*. The company has identified four internal performance strategies that it must pursue. The first two are 'provide ideal service' as it moves into new TAs and 'identify and fix delivery issues' in its existing business. Some potential metrics for these two strategies are:

1　Identify and fix delivery issues
 a.　Number of delivery issues encountered per project in the last three months – **Q**
 b.　Average time from discovery of each delivery issue to completed fix (for fixes completed during the last three months) – **CT**
 c.　Average hours expended per delivery issue fixed over the last three months – **E**
 d.　Percentage of projects on schedule during the last three months (as compared to forecast) – **T**

2　Provide ideal service
 a.　Percentage of active projects that have a set of 'ideal service' characteristics agreed to by customer and *CROmed* – **Q**
 b.　Percentage of 'ideal service' characteristics that *CROmed* is achieving at the end of the month on all projects – **Q**

Metrics 1a through 1d are classic performance metrics. *CROmed* is attempting to reduce the number of delivery issues, or improve performance when it does encounter an issue. Metrics are normalized to a quarter in order to provide consistency when comparing data over time. Note that in metric 2b, the fixes that are *completed* per quarter are the ones measured, rather than those *identified* per quarter. That definition allows *CROmed* to get actual data on each fix. The efficiency measure (1c) will require that employees accurately track their time and that the accounting department supply work codes that differentiate between normal project work and 'issue fix' work. This is a difficult change for employees to make, but it is critical if *CROmed* is to figure out whether things are really improving. Metric 1d simply makes sure that the organization isn't slowing its delivery pace while attempting to deal with its quality problems.

Meanwhile, metrics 2a and 2b focus strictly on quality and *CROmed*'s ability to develop a set of ideal service characteristics for each project. Since the concept of ideal service is new to *CROmed*, the first task is simply to get every project to embrace the idea (metric 2a) and then pay attention to those characteristics throughout the project (metric 2b). Once these metrics both are at or near 100 percent, *CROmed* can change the metrics to reflect other aspects of its ideal service model.

VirtuPharm

In *VirtuPharm*'s case, we want to develop metrics that measure the performance of the partnerships between *VirtuPharm* and its CROs. Possible partnership measures are:

1 Excellent project quality
   a. Percentage of documents requiring one or fewer review cycles for documents finalized in the last three months – **Q**
   b. Number of 'critical' site situations found by quality assurance (QA) during site audits – **Q**

2 On-budget performance and timely reimbursement
   a. Total billings per project milestone vs. original contract – **E**
   b. Business days from receipt of bill by *VirtuPharm* to receipt of payment by CRO – **CT**

3 On-time, short cycle time performance
   a. On-time contract signing – **T**
   b. Percentage of cycle times at or below forecast for milestones achieved in last three months – **CT**
   c. Percentage of milestones achieved on-time or early for milestones achieved in last three months – **T**

4 Active problem prevention and solving
   a. Average time from problem report to problem resolution for problems resolved in the last three months – **CT**
   b. Number of failure modes that were predicted at the beginning of each project and that had prevention strategies (for projects initiated in last three months) – **Q**

5 Accurate coordination and project planning
   a. Number of contract change orders – **Q**
   b. Number of forecast updates required after baseline forecast – **Q**

We've identified 11 metrics here with a heavy emphasis on quality. We could have emphasized cycle time, timeliness or efficiency further, had we so desired. However, the particular strategies identified for *VirtuPharm* lend themselves well to quality metrics. Metric 4b is an interesting one. The goal here is to encourage failure mode analysis and contingency planning at the beginning of the project. It's not strictly a quality metric, and it's not normalized. Once all projects are doing a good job of up-front failure analysis, this metric can be dropped.

Inventrix

Inventrix has perhaps the most difficult strategy of the four companies: it is attempting to build its product line through means other than discovery, and grow relationships with its customers on existing products. In this case, R & D is only one player in achieving growth through innovation. R & D must find ways to work with other parts of the company, especially sales and marketing. Here are some possible metrics *Inventrix* R & D could use for its first two internal performance strategies (build its product line through acquisition and through innovation). R & D built a simple process map for its innovation process, as shown in Figure 4.14, and then assigned appropriate metrics.

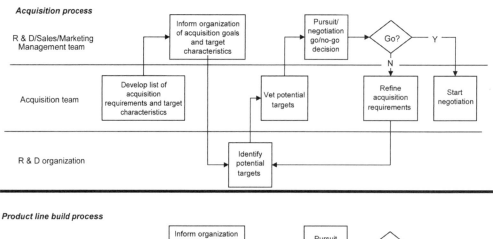

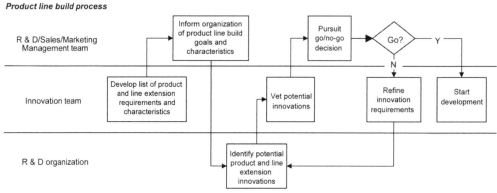

Figure 4.14 Process flows for two of *Inventrix*'s internal performance strategies

1 Build product line through acquisition
   a. Number of new acquisition targets suggested by R & D personnel in the last three months (vs. target of two acquisition targets per quarter) – **T**
   b. Weeks from identification of an acquisition target to go/no-go decision (for go/no-go decisions made in the last three months) – **CT**
   c. Percentage of R & D-identified acquisition targets that make it to negotiation stage (for negotiations begun in the last 12 months) – **Q**
   d. Percentage of employees that have spent at least 10 percent of their hours in the past three months on acquisition teams which include Sales and Marketing – **E**

2 Build product line through innovation
   a. Number of innovative products and line extensions suggested by R & D personnel in the last three months (vs. target of two products and four extensions per three-month period) – **T**
   b. Weeks from suggestion of an innovation to go/no-go decision on whether to pursue (for go/no-go decisions made in the last three months) – **CT**
   c. Percentage of R & D-identified innovations that receive a 'go' decision from management (for go/no-go decisions made in the last three months) – **Q**
   d. Percentage of employees that spend at least 10 percent of their hours on innovation teams in the last three months – **E**

SUMMARY

In this chapter, we've developed the techniques for creating and defining internal performance metrics. These are the easiest type of metrics for scientists to develop, since they are highly quantitative and precise. Hence, I suggest starting with these and then moving on to financial, customer and organizational growth metrics (which we'll discuss in Chapter 5).

We have defined six principles of internal performance metrics. They're listed in Figure 4.8 and repeated here as Figure 4.15 for your convenience. If you follow these principles during your metrics development, you'll have a much easier time.

I recommend that you do not have more than 12–18 total internal performance metrics. Otherwise, you'll just have too much to remember!

While many metrics experts rely on three types of performance metrics (Q, T, E), I like to add a fourth: CT. I believe there are parts of a process which are best measured by timeliness, while other parts are better measured by cycle time. Both metrics are derived from the same data, but the two types provide different information.

Principle 1:	Select metrics that can be categorized as timeliness, cycle time, quality or efficiency
Principle 2:	Measure what's important, not what's easy
	Principle 2a: Measure causes, not just effects
Principle 3:	Create a balanced set of metrics
	Principle 3a: Measure not only what you want to fix, but what you don't want to get worse
Principle 4:	Use process mapping to find internal performance metrics
	Principle 4a: Use internal, predictive metrics when possible
Principle 5:	Restrict your metrics to the critical few
Principle 6:	Carefully define your metrics
	Principle 6a: Normalize your metrics

Figure 4.15   Principles of performance metrics selection

CHAPTER 5

Rounding Out Your R & D Scorecard: Developing Financial, Customer and Organizational Growth Metrics

I was working with the regulatory affairs group in a mid-sized pharmaceutical company a few years ago. We were working on various process improvements and attempting to streamline their work flows. In the midst of our efforts, they went into hyper-drive to deliver a new drug application (NDA) to the US FDA. These folks worked night and day, seven days a week for a month to integrate the various reports from clinical, chemistry, manufacturing and controls (CMC), and so on, and to build a cohesive document. They did it, and the NDA was ultimately approved. Everyone congratulated themselves on a job well done under pressure (the fire fighters had triumphed again!). But I couldn't help wondering at the cost: several employees had left, admitting that the strain was too great. The regulatory affairs (RA) group was frustrated with CMC and clinical because those two groups had left it to RA to pull everything together and make the document look like a single, seamless whole. In essence, this was a huge example of suboptimization, with the last group in line (RA) having to pick up and assemble all of the pieces.

A year later, it turned out there was another chapter to this story. The new drug hit the market six months later than its competitor, and with a more limited label. The result was that sales were disappointing. Marketing was furious that R & D had not 'listened carefully' to its timing and labeling goals. Management was flummoxed as to why a supposedly successful development effort had yielded such poor results.

In my subsequent work with other pharmaceutical companies as well as CROs, I have noticed these same phenomena occur over and over again:

- Large turnover caused by overwhelming workloads
- Suboptimization by some groups resulting in extra work for downstream groups
- Frustration by downstream customer groups with the output of their predecessor supplier groups (see the discussion of NOAC in Chapter 4)
- Equivalent frustration by clinical concerning the lack of input from its customer groups (for example, marketing)
- Disappointing financial results
- Constant communication lapses and confusion about what to expect
- Eventual surprise and disappointment when things turn out poorly, even though much of it could have been predicted.

KEY METRICS CATEGORIES

Certainly, the internal performance metrics discussed in Chapter 4 are critical to improving this situation, but there are additional aspects to consider.

Financial

It's fine to turn out great products. But if every product loses money, the most robust company will eventually go bankrupt. And lest you think that this is all about money, even the most altruistic, not-for-profit organization has to at least break even financially if it is to survive. So, successful financial results are imperative regardless of the product being produced or the motivation of the organization.

Customer satisfaction

Whether your customer is the marketing and sales arm of your company or the physicians and patients who use your product, it's critical to make sure that the people who use the output of R & D will be happy with what they're getting. This may seem silly to some in R & D; after all, isn't R & D the engine that drives the company? So shouldn't R & D be the main decision maker in what gets developed? At first glance, these seem reasonable questions, and traditionally R & D has indeed been the main decision maker. However, there's no point in creating something that no one wants or which will never be used. Whether you're creating a drug to combat a common disease in the developed world or a specialty device to treat a rare, developing-world condition, the resulting product has to meet the medical, financial, logistical and social needs of the population that will use it. Developing a drug or device that doesn't meet all of these needs (often identified by a marketing or sales group) means an increased probability of failure. Even if the drug or device succeeds in the short run, customer dissatisfaction opens the door for competitors to supply something better. If the customer isn't satisfied, sooner or later you'll have problems.

Organizational growth

Organizational competencies and skill levels, employee satisfaction, employee career growth, leadership, work processes, technologies, information management, culture. A strong set of these capabilities at all levels of the organization are critical if your organization is to achieve real quantum leaps in performance. Organizational growth is perhaps the most nebulous category, but it underpins all of the other areas.

Just as we had to create balance between our internal performance metrics categories (timeliness, cycle time, quality and efficiency), we need to create balance between the four categories of internal performance, financial, customer satisfaction and organizational growth. Hence, the term 'balanced scorecard'. This multi-dimensional balance is shown in Figure 5.1.

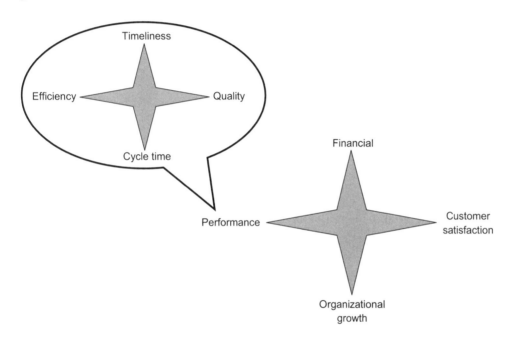

Figure 5.1   Good metrics systems maintain a balance in multiple dimensions: between financial, customer satisfaction, organizational growth and performance; and between timeliness, quality, cycle time and efficiency

Since we have already addressed our internal performance measures (see Chapter 4), we now have to ask and answer three additional questions:

1   How should we measure the organization's success in achieving its financial goals?
2   How can we measure whether customers are satisfied with what we produce?
3   How should we measure the organization's capabilities and competencies relative to those required to achieve its internal performance, customer satisfaction and financial goals?

Once we've done this, we will have developed a complete balanced scorecard.

It turns out that it is easiest to develop a scorecard by working through each of these areas in order. In the following sections, we'll deal with each area in detail, starting with financial metrics, then moving on to customer satisfaction and finally organizational growth metrics.

Before you begin, you may want to review the discussion of strategy maps in Chapter 3. Strategy maps are a particularly helpful tool in speeding the development of metrics in these three categories, and I highly recommend that you build your strategy map before trying to develop these metrics.

DEVELOPING FINANCIAL METRICS

Financial metrics attempt to address the question of whether we're making or losing money over time. At the top of an organization, financial metrics often measure overall revenue and profitability:

- Sales
- Earnings
- Profit
- Return on investment (ROI)
- Return on assets (ROA)
- Cost.

These types or measures are often not applicable to an R & D organization. First, there is no clear sales or profit to be measured. Certainly the organization as a whole can track its profit and loss. However, attempting to tie the profit of a drug or device in the marketplace to the effectiveness of the R & D organization is extremely difficult. A number of attempts have been made to link the cost of drug development to the three- or five-year sales of a drug, but that requires years of data after the drug hits the market, making the measure almost useless due to the long lag time. Similarly, people have tried to tie current sales or profit to current R & D spend. However, current sales are the result of long-past R & D investment and have no relation at all to current R & D spend. And none of these measures takes into account the confounding factors of the marketplace (such as, competitor efforts, changing regulatory climate, or intervening medical breakthroughs), which can render useless even the most effective R & D effort.

Second, cost can be measured, but it's not clear whether under-running your budget is good or bad. If a yearly budget under-run was caused by postponing a clinical trial that's critical to an NDA filing, that's not a good thing. Similarly, a yearly budget overrun that results from speeding up a critical trial could be a very good thing indeed! (This is an excellent example of the 'so what' test for good metrics. We'll discuss the 'so what' test and other metrics development rules in Chapter 6.)

Even though these traditional financial metrics don't make as much sense at the R & D level as they do at the corporate level, it's still important to measure what's going on financially. Some typical financial measures that are useful in an R & D environment are shown in Figure 5.2.

```
Actual clinical trial costs vs. budget
Actual development plan costs vs.
  budget
Sales from new products
Earned value
```

Figure 5.2 Typical R & D financial metrics

Actual costs vs. budget

A few paragraphs back, I mentioned that annual R & D budget overruns may not be a bad thing if they result in better products to market faster. Therefore, you may be wondering why 'actual vs. budget' metrics are listed in Figure 5.2. Here's my reasoning. Development and clinical trial budget overruns and under-runs almost always occur because of poor planning. By measuring and attempting to improve these metrics, we will in effect be improving our planning and forecasting. That in turn improves overall organizational performance. If we have a budget overrun on a project or trial, the money has to come from somewhere. R & D organizations usually have to starve some other project to get this one done. So better forecasting means more stable funding for all projects.

You may at this point be tempted to say that overruns occur because of unexpected events that overwhelm otherwise good plans; the planning was done well, but we couldn't have predicted the events that occurred later. I'd counter that any plan that doesn't account for most of these unexpected events is in effect a poor-quality plan. Using an example we've discussed previously, almost every trial runs into enrollment problems. You'd think that, given the huge experience base that the industry has with enrollment problems, we could at least accurately predict the enrollment schedule! In fact there are excellent, proven techniques out there for both accurately predicting and seriously improving enrollment. However, these techniques are rarely used and enrollment forecasting remains a dark art. The resulting schedule and cost overruns almost always could have been predicted had the planning been more realistic. The same is true with other parts of the trial or development: most budget problems come more from our unwillingness to plan and forecast than from our inability. Since 'what gets measured gets fixed', putting these types of metrics into your system will show your organization that accurate planning and forecasting is a serious goal.

Sales from new products

This is a metric that appeals to many organizations. While not requiring the long data collection cycle of a metric that measures profitability of certain drugs, it does give a sense of the value added by the R & D organization. Simply measuring the dollar contribution of drugs/devices that have been on the market less than three or four years, or the ratio of sales of these products to older products, gives a sense of the robustness of your company's portfolio and the contributions of R & D.

A variation on this metric is the projected sales of drugs and devices currently in the pipeline. This is typically expressed as the net present value of the first three to five years of sales for each product or the net present value of the next five calendar years of the products in the pipeline. Either way, this metric is strictly a forecast. Given the industry's lack of success in predicting which drugs will or will not become blockbusters, this metric seems to have little value in promoting anything beyond inflated sales forecasts.

Earned value

Perhaps the most valuable financial metric that can be used in R & D is earned value. EV has been used for many years in the construction and aerospace industries to measure financial and project progress in huge, complex undertakings such as building construction or new aircraft development. Since development of a new drug or device certainly qualifies as a huge, complex undertaking, it would seem reasonable to apply EV to these efforts. As it turns out, EV is a tremendously useful tool in drug and device development. Earned value provides a standardized way to:

- measure a project's progress (a clinical trial, drug development project or entire R & D portfolio)
- forecast its completion date and final cost
- provide schedule and budget variances
- provide consistent, numerical indicators for evaluating and comparing projects.

In the following paragraphs, I'll discuss in detail how to compute EV for a project. Basically, we're trying to compare the scheduled (or 'budgeted') work to the actual work performed at any given time. EV assumes that we (1) can create a complete schedule, as well as a budget that is tied to that schedule and (2) know the status of our schedule and costs (the 'actuals') at any given time during project execution. Knowing this information allows us to compare budget to actuals and thereby predict end-of-project performance. There's a bit of math involved and the terminology can be a bit confusing, but take your time and you'll figure it out!

EV can be computed at any point in the life of a project and requires five main pieces of data:

- **BAC – budget at completion.** This is the total, original, budgeted cost for the entire project.

- **CT – cycle time.** This is the scheduled time required to complete the entire project.

- **BCWS – budgeted cost of work scheduled.** At any point in time, the original forecast specifies a *specific amount of work that is to be done*. This may be more or less than the amount of work that has actually been done. BCWS is the total cost of the work that should have been accomplished so far. It answers the question: 'How much work should have been done?' BCWS can often be very hard to calculate, so it is often approximated by multiplying the total budget at completion (BAC) by the fraction of total project duration at the analysis date. For instance, on a one-year, $1 m project, the BCWS can be approximated at the end of month 4 as $333 000 (1 000 000 × 4/12).

- **BCWP – budgeted cost of work performed.** At any point in time after the start of a project, a specific amount of work has *actually been completed*. This amount of work may be more or less than was originally planned. Regardless, BCWP is the cost originally budgeted to accomplish the work that has been done so far. It answers the question: 'How much work was actually done?'

- **ACWP – actual cost of work performed.** At that same point in time, we also know (hopefully!) how much we've spent so far. This is ACWP – the actual cost of the work performed so far. This answers a third question: 'What did the work that was actually done actually cost?'

Knowing these five pieces of data, we can now calculate seven very useful measures of project status (Figure 5.3 shows the equations for each of these):

- **SV – schedule variance.** By subtracting the BCWS from the BCWP, we can figure out how far ahead or behind schedule we are. Positive SV means ahead of schedule (we've accomplished more than we'd planned), while negative SV means behind schedule (we've accomplished less than we'd planned). Positive numbers are good.

- **SPI – schedule performance index.** By dividing the BCWP by the BCWS, we can create the schedule performance index. A value of 1.0 means everything is right on schedule, while values above 1.0 mean ahead of schedule and values below 1.0 mean behind schedule. Values greater than 1.0 are good.

- **CV – cost variance.** By subtracting the ACWP from the BCWP, we can figure out how far ahead or behind budget we are. Positive CV means under budget (we've spent less than we'd planned for the work we've accomplished), while negative CV means over budget (we've spent more than we'd planned for the work we've accomplished). Positive numbers are good.

- **CPI – cost performance index.** By dividing the BCWP by the ACWP, we can create the cost performance index. A value of 1.0 means everything is right on budget, while values above 1.0 mean under budget and values below 1.0 mean over budget. Values greater than 1.0 are good.

- **EAC – estimate at completion.** By dividing the BAC by the CPI, we can create the EAC.

- **SAC – schedule at completion.** By dividing the scheduled project cycle time (CT) by the SPI, we can calculate a new overall project schedule. Thus, if the original schedule was 12 months and the SPI is 0.5 (that is, way behind schedule) the current projected schedule becomes 24 months from start to finish.

ROUNDING OUT
YOUR R & D
SCORECARD

- **VAC – variance at completion.** By subtracting the EAC from the BAC, we can calculate by how much we will over-run or under-run our original budget. Positive VAC means an under-run, while negative VAC means an overrun. Positive numbers are good.

Variable		Formula	Meaning
SV	Schedule variance	SV = BCWP – BCWS	< 0 = Behind schedule
SPI	Schedule performance index	SPI = BCWP / BCWS	< 1 = Behind schedule
CV	Cost variance	CV = BCWP – ACWP	< 0 = Over budget
CPI	Cost performance index	CPI = BCWP / ACWP	< 1 = Over budget
EAC	Estimate at completion	EAC = BAC / CPI	
SAC	Schedule at completion	SAC = CT/SPI	
VAC	Variance at completion	VAC = BAC – EAC	

Figure 5.3 Earned value calculations

Figure 5.4 gives an example of how earned value can be used to give a financial picture of a clinical trial enrollment. Let's assume we need to enroll 1000 subjects over 50 weeks,

At project completion:

Total enrollment	1000 subjects	
CT	50 weeks	(20 subjects/wk)
BAC	$500 000	($500/subject x 1000 subjects)

At week 10:

Forecast subjects enrolled	200	(20 subjects/wk x 10 wks)
Actual subjects enrolled	150	
BCWS	$100 000	(10 wks x 20 subjects/wk x $500/subject)
ACWP	$90 000	
BCWP	$75 000	(150 subjects x $500/subject)
SV	– $25 000	(BCWP – BCWS)
SPI	0.75	(BCWP/BCWS)
CV	– $15 000	(BCWP – ACWP)
CPI	0.83	(BCWP/ACWP)
At completion:		
EAC	$600 240	(BAC/CPI)
VAC	– $100 240	(BAC – EAC)
SAC	67 weeks	(CT/SPI)

Figure 5.4 Enrollment example of earned value analysis

which means a straight-line enrollment of 20 subjects. Let's further assume that it will cost us $500 per enrolled subject, resulting in a total enrollment cost of $500 000 (our BAC).

Now, let's look at a specific point in time during our project – week 10 of enrollment:

- At the end of week 10, our forecast says that we would have enrolled 200 subjects (20 subjects per week × 10 weeks) at a total cost of $100 000 (our BCWS).
- Let's say that, in actuality, we only enrolled 150 subjects by the end of week 10, and we spent $90 000 doing it (our ACWP).
- Meanwhile, we know that for the 150 subjects we enrolled, we should have spent $75 000 (our BCWP).

From these four values, we can calculate the various earned value measures shown in Figure 5.4. The SV of –$25 000 and SPI of 0.75 indicate that we are behind schedule, while the CV of –$15 000 and CPI of 0.83 indicate that we are over budget. Finally, our EAC of $600 240 and SAC of 66.7 indicate that – if nothing changes – we will need to come up with an extra $100 240 to complete our enrollment and it will require an additional 17 weeks.

This was, of course, a very simple example. The real power of earned value analysis (EVA) becomes evident when multiple parallel tasks and projects are tracked as a group. For instance, while it's easy to see the budget and schedule problems in this example, it's much more difficult to figure out the budget and schedule picture when enrollment, CMC, drug distribution, data management and other clinical trial functions are all running simultaneously – some ahead of schedule and budget and others behind. When we're running multiple trials simultaneously, the situation becomes even more complex. EVA allows you to integrate all of these different pieces and create an overall measure of what's happening across a trial, across a TA or even across all of R & D. It allows you to get an ongoing picture of budgets and schedules, early warning of pending problems, and end-of-project forecasts.

An example of a compound development project EVA is shown in Figure 5.5a. As you can see from the data table, trials B and E are on or ahead of schedule and/or budget. Trial B will come in right on budget but two months early, while trial E will be a month and a half late but will come in $25 000 under budget – assuming everything continues as it has so far. Meanwhile, trials A, C and D will be both late and over budget. Trial A is in the worst shape, with an expected overrun of $1.6 m and a trial length of 45 months instead of the forecast 12.5 months. If you look at the rightmost column, you'll see that the five trials should have completed in 24 months, but will require 45 months and an aggregate $1.1 m of excess funding.

The graph in Figure 5.5b provides an at-a-glance perspective on what is likely to happen. As of the June 1, 2005 analysis date (vertical line in the figure), the BCWP is well below the BCWS; that is, much less work has been accomplished than was originally planned. Meanwhile, the ACWP is higher than the BCWP, meaning that the work carried out so far is running over budget. The dashed line from the current ACWP to the EAC (which is literally off the chart) shows the expected trend assuming that the remainder of the project will proceed at the same rate as it has so far.

Analysis Date: 1-Jun-05

	Trial A		Trial B		Trial C		Trial D		Trial E		Overall performance
Forecast Data:											
Total Budget, BAC (US $)	$2 000 000		$4 000 000		$500 000		$1 000 000		$250 000		$7 750 000
	Milestone forecast dates	*Budget expenditure forecast*	*Milestone forecast dates*	*Budget expenditure forecast*	*Milestone forecast dates*	*Budget expenditure forecast*	*Milestone forecast dates*	*Budget expenditure forecast*	*Milestone forecast dates*	*Budget expenditure forecast*	
Start date	1-Jan-04	$0	1-Jul-04	$0	1-Oct-04	$0	1-Jan-05	$0	1-Feb-05	$0	
Protocol signed off	1-Feb-04	$200 000	15-Aug-04	$400 000	1-Nov-04	$50 000	1-Feb-05	$100 000	1-Mar-05	$25 000	
First patient first visit (FPFV)	15-Feb-04	$300 000	30-Aug-04	$600 000	15-Nov-04	$75 000	15-Feb-05	$150 000	15-Mar-05	$37 500	
Last patient first visit (LPFV)	15-Jun-04	$800 000	28-Feb-05	$1 600 000	15-Mar-05	$200 000	15-Jun-05	$400 000	15-Apr-05	$100 000	
Last patient last visit (LPLV)	15-Aug-04	$1 200 000	31-May-05	$2 400 000	15-Apr-05	$300 000	30-Jul-05	$600 000	15-Apr-05	$150 000	
Database closed (DBC)	1-Sep-04	$1 300 000	15-Jun-05	$2 600 000	30-Apr-05	$325 000	15-Aug-05	$650 000	1-May-05	$162 500	
Database locked (DBL)	15-Sep-04	$1 400 000	1-Jul-05	$2 800 000	15-May-05	$350 000	1-Sep-05	$700 000	15-May-05	$175 000	
Statistical analysis completed (SA)	15-Nov-04	$1 600 000	1-Sep-05	$3 200 000	15-Jul-05	$400 000	1-Nov-05	$800 000	15-Jul-05	$200 000	
Draft final report (FRD)	15-Dec-04	$1 800 000	30-Sep-05	$3 600 000	15-Aug-05	$450 000	30-Nov-05	$900 000	15-Aug-05	$225 000	
Signed-off final report (FR)	15-Jan-05	$2 000 000	31-Oct-05	$4 000 000	15-Sep-05	$500 000	31-Dec-05	$1 000 000	15-Sep-05	$250 000	
Project Status at Analysis Date:	50% of enrollment completed		Database locked		Last patient first visit		50% of enrollment completed		Last patient last visit		
EVA Metrics at Analysis Date:											
CT (months)	12.5		16		11.5		12		6.5		24
BCWS	$2 000 000 (trial complete)		$2 400 000 (LPLV)		$363 000 (1/4 SA)		$370,000 (7/8 enroll)		$181 000 (1/4 SA)		$5 314 000
BCWP	$550 000 (1/2 enroll)		$2 800 000 (DBL)		$200 000 (LPFV)		$275,000 (1/2 enroll)		$150 000 (LPLV)		$3 975 000
ACWP	$1 000 000		$2 800 000		$300 000		$300,000		$135 000		$4 535 000
SV	-$1 450 000		$400 000		-$163 000		-$95 000		-$31 000		-$1 339 000
SPI	0.28		1.17		0.55		0.74		0.83		0.75
CV	-$450 000		$0		-$100 000		-$25 000		$15 000		-$560 000
CPI	0.55		1.00		0.67		0.92		1.11		0.88
EAC	$3 636 364		$4 000 000		$750 000		$1 090 909		$225 000		$8 841 824
SAC	45		14		21		16		8		45
VAC	-$1 636 364		$0		-$250 000		-$90 909		$25 000		-$1 091 824

Figure 5.5a   Example of an earned value calculation for a set of five clinical trials

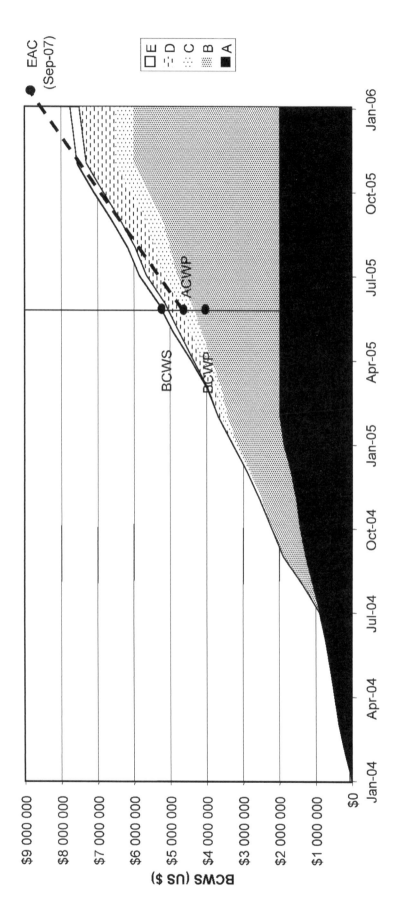

Figure 5.5b Example of an earned value graphical forecast for a set of five clinical trials

Armed with this information, management can now make intelligent decisions about how to either increase funding or otherwise work around the problem. Had the earned value been calculated from the start of trial A, management would have had an excellent way of monitoring what's going on.

Compared to EVA, most other financial measures pale – mainly due to the fact that they are retrospective and so provide very little value for management and decision making.

DEVELOPING CUSTOMER METRICS

The customer metrics that you develop for your organization come, of course, from the customer strategies that you developed in your strategy map. However, there are some good questions that you should ask as you're developing your customer metrics. Figure 5.6 lists some questions you may want to ask yourself, if you didn't ask them when you developed your strategy map in Chapter 3. Figure 5.7 lists some typical customer metrics.

- Who are your customers?
    - How should you measure their requirements?
    - How should you measure their satisfaction with what you produce?
- How do you build and maintain relationships with your customers?
    - How can you measure the quality of your relationships with them?
    - How can you measure how much your customers know about your operations, efforts and products?
- What are your competitors doing?
    - How can you measure how well you're doing compared to them?

Figure 5.6   Some useful customer metrics questions

- Customer satisfaction with your 'products' (for example, the appropriateness of your pipeline vis-à-vis the needs of your marketing/sales organization, as well as physicians and patients).
- Customer view of our company compared to competitors (for example, if you're using a product leadership strategy, do physicians and patients look at your company that way?).
- Customer loyalty to your company and its products.
- Market share in key therapeutic areas.
- Annual growth in market share in key therapeutic areas.

Figure 5.7   Typical customer metrics

Depending on which customer strategy you plan to use (refer to Figures 3.4 and 3.5 for a discussion of strategy options), your specific metrics will vary. For instance, a customer intimacy strategy may prompt you to measure physician and patient loyalty, whereas a product leadership strategy may mean that you want to measure physician and patient perceptions of your company as a maker of first-in-class or best-in-class drugs or devices. Regardless, customer measures often tend to be survey-based. However, you can also measure customer reaction by tracking your market share in a particular indication or your annual market share growth. Ideally, a combination of survey responses and hard data should be used.

You may also want to include supplier metrics in this category. Your suppliers aren't typically your customers, but their satisfaction is critical to your long-term success. Some questions to ask about suppliers are listed in Figure 5.8.

- How do you select suppliers?
  - How consistent are you in your selection process?

- How effective are you at working with your suppliers?
  - How satisfied are your suppliers with you as a customer?
  - How satisfied are your teams with their suppliers?
  - How good are your communication and coordination processes?

- Are you and your suppliers both prospering from the relationship?
  - Are your costs down and quality up?
  - Is their profit up?
  - Are you giving them repeat business?

Figure 5.8 Some useful supplier metrics questions

DEVELOPING ORGANIZATIONAL GROWTH METRICS

Having developed internal performance metrics in Chapter 4 and financial and customer metrics earlier in this chapter, it's time to look at the last – and most difficult – category: organizational growth metrics. There are really five components of organizational performance:

1 Competencies – how good is our organization at doing what we need to do?
2 Technologies and processes – do we have state-of-the-art capabilities where we need them?
3 Information management – can everyone get to the information they need when they need it?
4 Employees – are employees satisfied, motivated and growing?
5 Organization – is the organizational culture appropriate and is there sufficient leadership and work management?

Figure 5.9 lists possible questions to ask yourself when developing organizational growth metrics. There are many others that will likely result from your scrutiny of your strategy map, but these questions may help to focus your thinking.

The biggest trick in developing organizational growth metrics is to narrow them down to a few key metrics that will actually measure and reflect progress by the organization. Too

- **Competencies**
  - Number of competencies available compared to required
  - Number of employees with appropriate type and level of skills
  - Ability to innovate and respond quickly to challenges

- **Technologies and processes**
  - State-of-the-art technologies wherever required
  - Percent of processes that are effective and efficient (esp. key value creation and support processes)
  - Percent of key processes with improvement plans and targets
  - Emphasis on effective (vs. ineffective) work

- **Information management**
  - Quality of the IM infrastructure
  - Percent of IM systems that are state-of-the-art
  - Deployment of key systems across the organization
  - Accessibility by employees

- **Employees**
  - Employee satisfaction and motivation
  - Quality of work environment
  - Hiring and career progression alignment to goals
  - Appropriate education and training
  - Feedback and performance management system effectiveness

- **Organization**
  - Strong culture aligned with organizational vision
    - Percent of organization that is aware of and accepts the key organizational goals
    - Social responsibility and ethical behavior
  - Leadership
    - Clear direction-setting and communication
    - Continuous review and feedback
    - Empowerment of the organization
  - Work management
    - Use of teams
    - Empowerment, innovation and agility

Figure 5.9   Potential organizational growth metrics questions

often, organizations measure lots of things that don't really track organizational progress. For instance, it's popular to measure the number of improvement projects underway each year across an organization. Management will set targets for the number of projects to be started and – since what gets measured does indeed get fixed – the organization will dutifully start at least this number of improvement projects. Whether these are the best projects to invest in or whether they are successful in improving the organization is never measured, and these projects rarely have any significant impact. On the other hand, measuring the number of improvement projects completed in a few key improvement areas has a much better chance of driving high-impact projects. If this metric is further refined by measuring the organizational impact of these projects (for example, dollars saved, quality improved, cycle times reduced), it has the potential to really drive focused organizational improvement.

CASE STUDIES

The following Figures (5.10–5.15) show a potential set of metrics for our example companies. Since *BioStart* had decided not to pursue a balanced scorecard, I've left them out of this discussion. However, a full scorecard could easily be produced for *BioStart* and I'll leave that to you as an exercise.

The metrics that follow are based on the strategy maps that were developed in Chapter 3. Those maps are included again here for ease of reading. You may find that you disagree with some of the metrics listed, or that you think others should be added to the list. That's just fine. I find that each organization has a different perspective on which metrics are important, even if the problems they are facing are the same. Since the goal of metrics is to help you improve *your* organization's performance, there's really no 'right' or 'wrong' set of metrics.

SUMMARY

Creating a full, balanced set of metrics is an exercise in system optimization. We are attempting to root out and eliminate suboptimization within the organization, ensure that everyone is focused on a relatively few key issues, and that we're producing results which are prized by our customers. In this chapter, questions have been provided which you can use to build metrics for the financial, customer and organizational aspects of your system. By asking these questions in concert with the strategy map developed in Chapter 3, you should have relatively little trouble developing a clear set of metrics.

I recommend that you have no more than 18 or 20 metrics for your entire system. While you may want more, remember that:

1 It's hard for the organization to keep too many metrics in the front of their collective brain.
2 The goal of these metrics is to focus attention, so the more metrics you have, the less the ability to focus on the 'critical few' things.
3 Organizational alignment is also very important. It's better to have alignment around only some of the critical issues, than to try to address all of them and lose alignment between groups and departments.

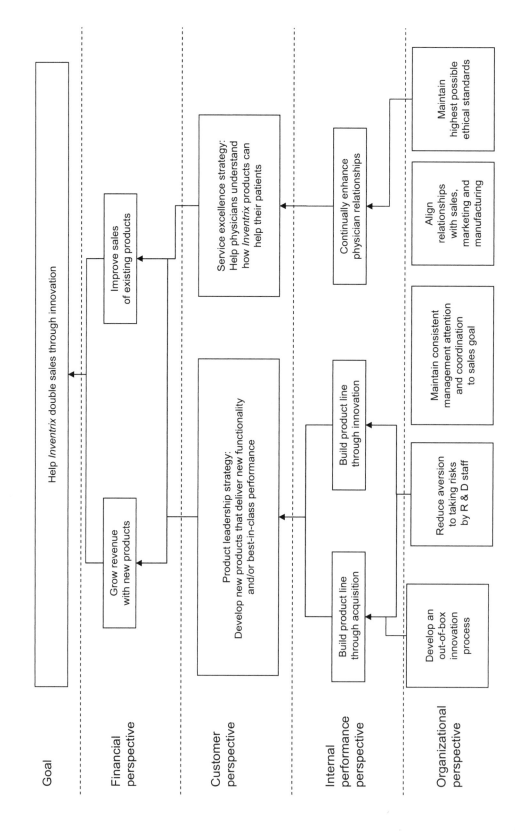

Figure 5.10 A potential *Inventrix* strategy map

Axis	Strategy	Metrics
Internal performance	Build product line through acquisition	1. Number of new acquisition targets suggested by R & D personnel in the last three months – **T** 2. Weeks from identification of an acquisition target to go/no-go decision on whether to begin negotiation – **CT** 3. % of R & D-identified acquisition targets that make it to negotiation stage – **Q** 4. % of employees that have spent at least 10% of their hours in the past three months on acquisition teams – **E**
	Build product line through innovation	1. Number of innovative products and line extensions suggested by R & D personnel in the last three months – **T** 2. Weeks from suggestion of an innovation to go/no-go decision on whether to pursue – **CT** 3. % of R & D-identified innovations that receive a 'go' decision from management – **Q** 4. % of employees that spend at least 10% of their hours on innovation teams – **E**
Financial	Grow revenue with new products (products on the market for less than five years)	1. % of annual sales from new products 2. % increase in new product sales over previous year
	Improve sales of existing products	1. % increase in existing product sales over previous year
Customer	New products/ functionality	1. Number of new product ideas that are best-in-class
	Help physicians	1. Number of thought leaders contacted in each TA 2. Number of position papers issued 3. Number of journal papers published
Organizational growth	Out-of-box innovation process	1. % of employees spending at least 10% of their time on innovation 2. Number of innovation suggestions
	Reduce risk aversion	1. Number of quarterly rewards given for risk-taking
	Maintain consistent management attention	1. % of management which include discussion of $10 m goal activities, status and action items
	Align relations with sales/marketing	1. Number of meetings on innovation which had senior R & D and marketing attendees
	Maintain highest ethical standards	1. % of employees saying on survey that *Inventrix* encourages and rewards ethical behavior

Figure 5.11 Potential *Inventrix* balanced scorecard metrics

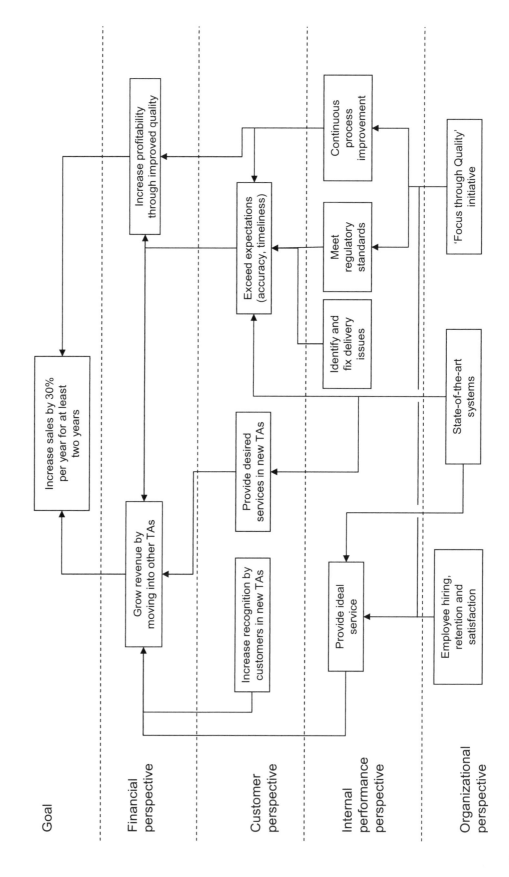

Figure 5.12 A potential *CROmed* strategy map

Axis	Strategy	Metrics
Internal performance	Identify and fix delivery issues	1. Number of delivery issues encountered per project in the last three months – **Q** 2. Average time from discovery of each delivery issue to completed fix (for fixes completed during the last three months) – **CT** 3. Average hours expended per delivery issue fixed over the last three months – **E** 4. % of projects on schedule during the last three months (as compared to forecast) – **T**
	Provide ideal service	1. % of active projects that have a set of 'ideal service' characteristics agreed to by customer and *CROmed* – **Q** 2. % of 'ideal service' characteristics that *CROmed* is achieving at end of the month on all projects – **Q**
Financial	Grow revenue through new TAs	1. % of sales coming from new TAs in the last 12 months 2. Annual sales by TA
	Increase profitability	1. Cost per work unit produced normalized to 2004 2. Quarterly profit as a percentage of sales
Customer	Increase recognition in new TAs	1. Number of new contacts made in new TAs 2. Percentage of existing pharma customers with pending contracts in new TAs
	Provide desired services in new TAs	1. Percentage of services identified on new customer surveys that are now available
	Exceed expectations	1. % of customer survey responses of 4 or 5 (on 5-point scale) indicating very or extremely satisfied with data accuracy and project timeliness
Organizational growth	Employee hiring, retention, satisfaction	1. 90% of new hires rate a 4 or 5 (on 5-point scale) relative to the competencies and qualifications required for the job using the PerfectHire rating scale 2. Employee voluntary turnover rate (annual) 3. % of employees giving *CROmed* an overall satisfaction score of 4 or 5 (on a 5-point scale) on the annual Employee Satisfaction Survey
	State of the art systems	1. % of 'mission critical' systems that are currently state of the art
	'Focus through Quality'	1. % of executive management meeting time spent on 'Focus through Quality' discussions 2. Number of 'Town Hall' meetings per quarter dedicated to 'Focus through Quality' initiatives

Figure 5.13   Potential *CROmed* balanced scorecard metrics

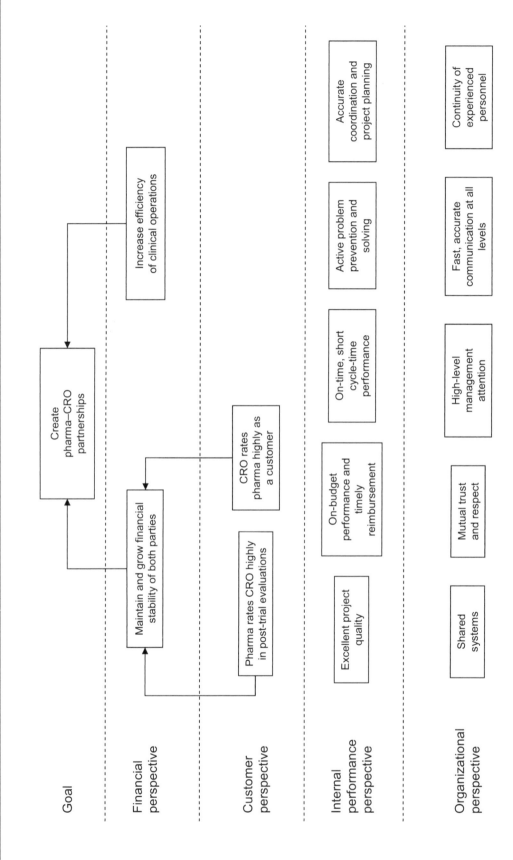

Figure 5.14 A potential *VirtuPharm* strategy map

Axis	Strategy	Metrics
Internal performance	Excellent project quality	1. % of documents requiring one or fewer review cycles for documents finalized in the last three months – **Q** 2. Number of 'critical' site situations found by QA during site audits – **Q**
	On-budget performance and timely reimbursement	1. Total billings per project milestone vs. original contract – **E** 2. Business days from receipt of bill by *VirtuPharm* to receipt of payment by CRO – **CT**
	On-time, short cycle time performance	1. On-time contract signing – **T** 2. % of cycle times at or below forecast for milestones achieved in last three months – **CT** 3. % of milestones achieved on-time or early for milestones achieved in last three months – **T**
	Active problem prevention and solving	1. Average time from problem report to problem resolution for problems resolved in the last three months – **CT** 2. Number of failure modes that were predicted using a Failure Modes Effects Analysis at the beginning of each project which also had prevention strategies (for projects initiated in last three months) – **Q**
	Accurate coordination and project planning	1. Number of contract change orders – **Q** 2. Number of forecast updates required after baseline forecast – **Q**
Financial	Financial stability of both parties	1. Clinical operations cost (Phases I though IIIb) per patient 2. Average % of 'reserve' funding consumed per trial 3. Average % of bonus fund that CRO receives due to budgetary efficiencies
	Clinical operations efficiency	1. Cost per unit of clinical operations work produced (including both *VirtuPharm* and CRO costs)
Customer	High pharma ratings	1. % of *VirtuPharm* mid-project and post-project survey responses of 4 or 5 (on 5-point scale) indicating very or extremely satisfied with CRO performance (survey administered by third party, results blinded)
	High CRO ratings	1. % of CRO mid-project and post-project survey responses of 4 or 5 (on 5-point scale) indicating very or extremely satisfied with *VirtuPharm* performance (survey administered by third party, results blinded)
Organizational growth	Shared systems	1. % of 'mission critical' systems, software and processes which are documented and shared between *VirtuPharm* and its CROs
	Respect and trust	1. % of *VirtuPharm* and CRO mid-project and post-project survey responses of 4 or 5 (on a 5-point scale) indicating high or very high level of respect and trust (survey administered by third party, results blinded)
	Management attention	1. Number of *VirtuPharm*-CRO coordination meetings per year in which Vice Presidents or above attend from both the CRO and *VirtuPharm*
	Communication	1. % of *VirtuPharm* and CRO mid-project and post-project survey responses of 4 or 5 (on a 5-point scale) indicating high or very high effectiveness of communication (survey administered by third party, results blinded)
	Personnel continuity	1. Turnover rate of core *VirtuPharm* and CRO personnel on *VirtuPharm*-CRO teams

Figure 5.15 Potential *VirtuPharm* balanced scorecard metrics

CHAPTER 6

Metrics Tips and Tricks

M any years ago, when metrics systems were in their infancy, I was working with a manufacturing client to build one of the first corporate-wide performance metrics systems. This client had multiple locations throughout the world and received a large number of complex parts from suppliers, which they then assembled into completed products. One of the big problems they were having was that there were lots of defects in the parts that suppliers sent to them. These defective parts were usually found by the incoming inspection group, but any that slipped through caused havoc in their assembly process. It seemed obvious that they should be measuring and improving the quality of incoming parts. So we developed what we thought would be a straightforward metric:

number of parts rejected by incoming inspection at a specific location

Of course, some locations received more parts than others, so we had to normalize the metric in order to compare across locations. We did this by writing the metric this way:

% of total incoming parts per month at a specific location
that were rejected by incoming inspection

All locations used a standard 'reject tag' to identify rejected parts, so we figured that we could just add up the reject tags at each location and divide by the number of incoming parts. This seemed both simple and straightforward. This fact in itself should have been a clue that we were headed for trouble!

So off we went to pilot the metric. We tried it at two locations and immediately ran into a problem:

- Location A did a 100 percent inspection of all parts and wrote a reject tag for every bad part, while
- Location B sampled parts from each shipment and wrote a single reject tag for the whole shipment if a bad part was found.

The result was that Location A always had more reject tags than Location B. We fixed this problem by asking Location A to issue a single reject tag for an entire shipment no matter how many bad parts were found in that shipment. They agreed and that changed our metric to:

% of total incoming shipments per month at a specific location that had reject tags

Having solved that discrepancy, we thought we had the metric pretty well defined. So we deployed it to all locations. Over the next few months we kept an eye on the data and found two interesting things:

1   Even though we hadn't set any goal for the metric, the reject tag percentage began a slow, steady decline. What was getting measured was indeed getting fixed!
2   Location C had a reject percentage that was less than half of any other location. It seemed as if Location C was certainly doing something right to get such high quality!

We went out to various locations to try to figure out what was happening. First we tried to understand the steadily decreasing reject tag percentages. It turned out that as soon as we started measuring corporate-wide, the various incoming inspection departments got a bit nervous, and very competitive. They started scrutinizing their own inspection processes to make sure a part really was bad before they rejected it (previously it was easier to reject a suspicious part and let the supplier fix it than to make sure that the part really wasn't usable). They also started calling problem suppliers and warning them that the corporation 'was watching'. This prompted the various suppliers to redouble their own inspection and quality efforts. The net result was that rejections slowly decreased, simply as a result of someone paying attention! As I've said already many times in this book, 'what gets measured really does get fixed'.

The second issue had a more surprising explanation. We went to Location C to see what had made their reject tag percentages so low. As you might have guessed, it wasn't because they had higher-quality suppliers. Instead it was due to the fact that Location C inspectors immediately called each supplier when a bad part or shipment was found. They then gave the supplier a chance to rework the shipment 'off the books'. Only if the supplier couldn't pass inspection after several tries did they issue a reject tag! In essence, they were circumventing the reject tag procedure in order to make their numbers look good. As with the first issue, what gets measured does indeed get fixed – just not always in the way you'd like!

By this point you've probably spent some time developing metrics of your own. (If you haven't, now would be a good time to experiment a bit.) There's a lot of art as well as science to developing the right metrics; and a lot of trial and error as well. Over the years, I've learned a lot about which types of metrics work – and which types don't. So in this chapter I'd like to go through a typical metrics development effort and give you some tips on what to do and what not to do in developing your own system.

Figure 6.1 provides a simple flow for developing your metrics set. As was discussed in earlier chapters, your metrics have to be tied to important goals in the organization. In a vision-driven metrics system, the goal is to track progress toward some future-state vision for the organization. In that case, you proceed down the left leg of the flow in Figure 6.1. The best approach is to use the strategy map process described in Chapter 3 – or some equivalent process – to develop a specific set of strategies for attaining the vision. Metrics development becomes fairly straightforward once these strategies are developed.

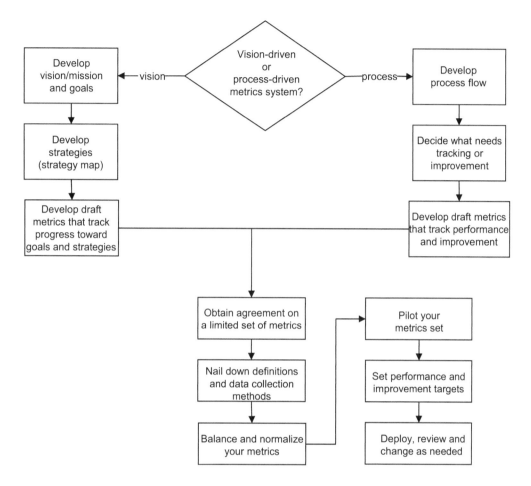

Figure 6.1   A decision tree for developing your metrics system

In a process-driven metrics system, the goal is to improve a specific business process in the organization, such as improving clinical trials or getting higher-quality go/no-go decisions during compound development. In this case, you proceed down the right leg of Figure 6.1. It's best to start with a map of the process (a 'process flow'), identify areas for improvement, and then attach metrics to those improvement areas.

Here's an example: *BioStart* wants to track and improve clinical trials performance. They would follow the 'process-driven' leg of Figure 6.1. They would be best off starting with a process map of their clinical trials process, such as the one shown in Figure 6.2. They could then identify specific areas that need to be tracked and perhaps improved. These areas might include better forecasting and execution of enrollment or techniques for reducing site queries, as shown in the figure. They would then develop internal performance metrics as described in Chapter 4.

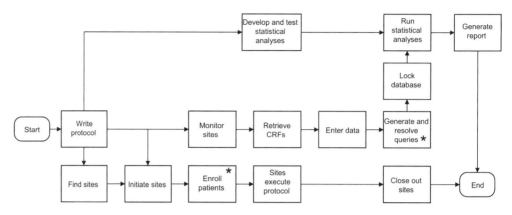

* Process problem identified by *BioStart* Clinical Operations Group

Figure 6.2 Example clinical trial process map with process issues highlighted

Meanwhile, *Inventrix* is attempting to improve the functioning of their entire organization and achieve a future-state vision of a highly innovative R & D operation. They would follow the 'vision-driven' leg of Figure 6.1. As described in Chapters 3–5, they would develop a strategy map and then develop balanced scorecard metrics to reflect their strategies and goals.

In either case, it's important to base your metrics development on a clear set of organizational or process goals. It's all too easy to simply start generating metrics, but that way madness lies. You'll end up with a huge pile of disconnected metrics, a data collection nightmare, and no improvement to show for it.

Metrics tip 1: Start at the top, either with a vision or a process and a goal. Don't fall prey to the temptation to generate metrics in a vacuum.

So far, so good. But now we need to develop a complete set of metrics that can be successfully deployed across the organization. As was found in the case study at the beginning of this chapter, that can be more difficult than it might seem. As you try to pilot and deploy your metrics, you'll end up disqualifying one metric after another:

● Some will be too expensive to collect.
● For others it will take too long to acquire the data, so they won't provide timely information.
● Still others will be difficult to normalize across trials, projects or therapeutic areas, so it will take a long time to gather enough data points.
● A few will look very good but be unacceptable to one part of the organization or another for any number of reasons.

As an example, one of my clients wanted to measure the performance of their CROs, but insisted that any metrics we developed use only existing data. No new data collection was allowed. We generated an 'ideal' set of metrics (those that we really wanted to use) and then proceeded to modify each or find substitutes to meet the 'no new data' constraint. It took us close to 100 metrics to find the 20 or so that we wanted.

While you may not have such a restrictive condition levied on your metrics development effort, you undoubtedly will have restrictions to contend with. You'll need lots of 'alternative' metrics on hand in case your preferred metrics become unacceptable for one reason or another. To generate this list of alternative metrics, interview everyone who has a stake in the goal and ask each how they would measure success. Each will give you a different perspective and usually a new metric or two. Then, from your larger list, select an 'ideal' set and ask this same group of interviewees to review and approve it. If these people have a stake in the goal, they'd better also have a stake in the metrics that they'll be measured by. Since you're looking for support for your system, make sure to interview a variety of people, including senior executives, middle managers, line managers, operating staff, department heads and anyone who might have to supply your data (such as clinical team leads). You'll eventually need support from all of them to make the system succeed.

Metrics tip 2: Get everyone who has a stake in the process or goal involved in metrics identification and selection. Include anyone who may have to contribute data to your system. Keep executives involved throughout.

Metrics tip 3: Generate a large list of potential metrics to choose from. You may need alternative metrics if some of your preferred metrics don't pan out.

At this point, it's important to ask a simple question: 'So what?' If I measure a particular parameter and collect data, will the result (1) point to an appropriate action and (2) drive the organization to take that action (that is, change their behavior)? R & D organizations collect hordes of metrics data that doesn't pass the 'so what' test. For example, it is standard to collect data on enrollment or number of queries, but those metrics are – by themselves – useless. Simply knowing the number of queries doesn't do any good, unless you also know how many were forecast at a given point in time. 'Number of queries relative to forecast' passes the 'so what' test because a high number tells me there's a problem that I'd better fix if I'm to come in on time and on budget, while a low number tells me either I'm doing something right, my forecasting isn't accurate or there's a problem lurking somewhere that I'd better be prepared for later.

Metrics tip 4: Make sure each metric passes the 'so what' test.

Once you've got an ideal set, it's time to nail down the metrics definitions. As you saw in the inspection example earlier in the chapter, it may take you several tries to get each definition right. This is pretty tedious work, but it's critical to ultimate success. The sooner you get this step done the better, so do it early. Here are the key questions you should ask when defining your metrics:

● How am I defining the measure itself? For instance, let's look at the cycle time from last patient last visit to database lock (LPLV-DBL). In LPLV-DBL, everyone defines DBL differently. Some say it's the date when the 'database is locked' memo is issued by data management. Others say that's not right, because that doesn't take into account the times (unfortunately all too frequent) when the database has to be 'unlocked' to correct errors. This faction argues that DBL is the point where the database is locked once and for all (hard to determine until after the fact). Still others argue that a better end point for this metric is the point at which all queries are resolved. In fact, it doesn't really matter which definition you use as long as everyone in your organization agrees to abide by that definition.

- How am I going to normalize the measure? Converting the value to a percent of total is often a good way to do this. Other ways are to divide the metric by some 'standard' value. In the example at the beginning of the chapter, we converted the raw rejection rate to a percentage of shipments for each location. That allowed us to compare across locations, which we couldn't do with the raw reject tag numbers. Similarly, measuring the number of queries resulting from a protocol might be a useful quality measure, but there's no way of comparing a huge but simple anti-hypertensive trial to a small, complex oncology trial. To allow comparisons, we have to normalize for size and complexity. We could use 'number of patients' as a representation of size and 'number of CRF pages' or 'number of safety and efficacy parameters' as a measure of protocol complexity. So a normalized query metric might look like this:

$$\frac{Number\ of\ queries}{(Number\ of\ patients) \times (Number\ of\ CRF\ pages)}$$

- What are the data sources for my metric? It's important early on to figure out where the data will come from. Are your teams or departments going to generate it? Can you get it from your CRO? Can you get it from your sites? Once you've decided who will provide the data, ask them whether they can actually provide the data as you've specified it. You'd be surprised that some seemingly simple data can actually be quite hard to obtain (and vice versa). An example of this is a normalization factor one client wanted to use. In attempting to normalize across therapeutic areas, they looked at several variables and finally settled on 'number of data points' as a good measure of trial size and complexity. It seemed that data management should be able to easily generate this value by adding up the number of cells in the database. However, there turned out to be so many factors involved in calculating this value that they had to abandon it and go with a different normalization factor.

- How often will data be collected? This question is often very difficult to answer. It's important to acquire the data often enough to be timely and useful, but the cost of data collection is directly proportional to collection frequency. Also, there's no point in collecting data on a daily basis when the process only generates a few pieces of data a week. You're going to have to make a tradeoff on each metric between data 'freshness' and collection cost. The best bet here is to let the goal for each metric drive collection frequency. And remember, all metrics don't have to be collected with the same frequency. You could collect some data weekly, while you collect other data quarterly or annually.

- What is my data collection method? Along with data collection frequency, you need to understand how the data will be supplied and collected. It's obviously easiest in the long run to automate as much as possible, but don't be bedazzled by automation. There are several top-quality metrics databases on the market that can be customized, automated, web enabled and more. That's great if you need a sophisticated system.

Many times, however, it's easiest to collect a small amount of data manually and simply enter it into your metrics system on a periodic basis. For instance, in the LPLV-DBL metric discussed above, we asked that the administrative assistant in charge of the metrics system be included on the 'database is locked' memo distribution list. Then, every time the assistant received one of these memos, she simply placed it in a 'pending metrics' file. Once a month she entered all of the memo dates in the database under the appropriate protocols. Tracking that data took less than 15 minutes a month. On the other hand, it's possible to have all of the data entered by the data owners through a website. This makes data entry easy, but puts pressure on the database administrator (see the next bullet).

● Who will administer the metrics database? The amount of work to maintain the database is directly related to the complexity of the metrics. If you've got lots of data from disparate sources, it may be easiest to have the data owner enter the data into a website. That would seem to make data entry easy, but someone has to maintain the website and database, as well as hound those laggards who never enter their data. On the other hand, even in this age of networked systems, sometimes it is still easiest to have one person who periodically collects and enters the data on a stand-alone computer. This may seem archaic, but it makes maintaining and administering the database quite simple.

● Who will receive what reports? Once you're collecting data, all sorts of people will ask you for information. It's wise to create a few simple reports that are available to everyone and provide custom reports only when necessary. That minimizes the workload and the volume of paper. (Reporting and display will be discussed further in Chapter 7.)

Metrics tip 5: Carefully define each metric, including definition of the measure, normalization factors, data sources, collection frequency and collection sources. Also, make sure you have a handle on who will provide database administration, who will receive reports and what level of report customization you'll need.

It is also important, as we've emphasized in previous chapters, to make sure your metrics are balanced. There are two aspects to this:

1 Balance your internal performance metrics between quality, efficiency, timeliness and cycle time (Chapter 4) and between financial, customer, internal performance and organizational growth (Chapter 5).

2 Balance your metrics between predictive (or prospective) and outcome (or retrospective) metrics.

If you find you have too many (or too few) of a specific metric type, remove (or create) some to make your set balanced. Also, make sure you don't have only retrospective metrics. These are by far the easiest to create, but often aren't as useful as prospective metrics which provide you with a glimpse of what's ahead.

Metrics tip 6: Create balance in your metrics system, both between metrics categories and between prospective and retrospective metrics.

When all is said and done, you're likely to end up with a pretty big set of metrics. Many experts argue that it's fine to have 30–40 metrics. I'm not one of those people. I find that I have trouble keeping more than two or three things in my brain at once, so there's absolutely no chance of my ever being able to remember 30 metrics! In my view, the goal of metrics is to help the organization improve, and it's best if everyone can keep the important improvement efforts in mind as they're going about their daily work and as they're making decisions. That means that you have to keep the number of metrics down to a critical few. Of course, building a balanced scorecard with only three to four metrics in total is unreasonable. That would mean only one metric per category. So I encourage a compromise: try to aim for a dozen metrics (three per performance category or balanced scorecard category), but never exceed about 18. In Chapter 8 we'll talk about metrics system flow-down, which is where you can use all of those other metrics that you can't bear to stop collecting.

The added advantage of limiting your metrics to the critical few is that it forces you to decide what's really important, and signals that to everyone in the organization. Too many metrics allows the organization to lose focus, while a critical few keeps everyone focused.

Metrics tip 7: Limit your metrics system to 12–18 metrics.

OK. You've finally got a limited, balanced set of metrics that are well defined, pass the 'so what' test, and have widespread backing from the organization. Are we done yet? Not quite. You still need to pilot the metrics set to make sure you can, in fact, collect the data you want and the metrics do indeed measure what you thought they would measure. As you may recall from the beginning example in this chapter, the rejection tag metric looked fine, but it broke down when it was applied because some parts of the organization tried to outsmart it. These are things that are best found out in a pilot phase rather than after full deployment. I generally find that you need three to six months to pilot a system before you know whether everything works as advertised.

Metrics tip 8: Pilot your metrics in one part of the organization for three to six months before full deployment.

Having piloted to make sure everything works right, now reconvene your original review and buy-in group (see metrics tip 2) and set performance targets for each metric. The pilot data will have given you a baseline for current performance, so you have a foundation for your targets. Make sure to set stretch goals (goals which force the organization to stretch), but don't make them impossible. Goal-setting is more art than science, so have your group identify 'business as usual' goals and 'pipedream' goals. Then set your target somewhere in between the two. For instance, if your cycle time from last patient last visit to database lock is normally 60 calendar days:

- a 'business as usual' goal might be 55 calendar days, since most people have little trouble improving processes by 10 percent

- a 'pipedream' goal might be 10 calendar days. If you're now at 60 days, it's ludicrous to think about cutting the time by a factor of 6 (at least in the near term)
- a reasonable goal would be somewhere between 10 and 55 days.

Having narrowed the range somewhat, you can pick the mid point of 32 days ([10+55]/2). If your team panics, then raise the goal a bit. When people start saying 'that goal is possible but I have no idea how to achieve it', you know you've got a stretch goal.

Metrics tip 9: Set realistic, stretch goals for each metric. Make them tougher than 'business as usual' but avoid 'pipedreams'.

At long last, it's time to deploy the system. Everything should work well, but there are still possible traps, so make sure to stay vigilant for odd trends in the data (see the example at the beginning of the chapter). Also, don't be afraid to adjust or change your metrics if you find that some of them are no longer useful. If your organization is making improvements, some metrics that were important early on will become unnecessary because you've made all the improvements that were possible in that area.

Metrics tip 10: After deployment, remain vigilant for data trends that might mean data isn't being collected properly or a metric has become obsolete.

Finally, don't bury your data in a computer! It's fine to use one of the many metrics collection and display systems out there. In fact, I encourage my clients to use these systems. However, the real impact of your system will result from prominent display of the data. Put it up on a wall. Go through the data at all your management meetings and employee 'town hall' meetings. The goal is to make the data a part of everyone's day-to-day consciousness so that they are always on the lookout for improvements. That won't happen if people have to look actively for the data in their computer. Metrics display systems will be discussed in Chapter 7.

Metrics tip 11: Display metrics data prominently and emphasize it at every meeting so that everyone in the organization is constantly aware of its importance and implications.

SUMMARY

Making sure you have the best metrics for your purpose, and that those metrics are properly defined and deployed, is a difficult task. In this chapter, we've discussed a systematic way of doing that and provided a series of tips to help you be successful. The process for developing and deploying your metrics is shown in Figure 6.1 (repeated here, as Figure 6.4 for your convenience), and the 11 tips are shown in Figure 6.3.

1. Start at the top, either with a vision or a process and a goal. Don't fall prey to the temptation to generate metrics in a vacuum.

2. Get everyone who has a stake in the process or goal involved in metrics identification and selection. Include anyone who may have to contribute data to your system. Keep executives involved throughout.

3. Generate a large list of potential metrics to choose from. You may need alternative metrics if some of your preferred metrics don't pan out.

4. Make sure each metric passes the 'so what' test.

5. Carefully define each metric, including definition of the measure, normalization factors, data sources, collection frequency and collection sources. Also make sure you have a handle on who will provide database administration, who will receive reports and what level of report customization you'll need.

6. Create balance in your metrics system, both between metrics categories and between prospective and retrospective metrics.

7. Limit your metrics system to 12–18 metrics.

8. Pilot your metrics in one part of the organization for three to six months before full deployment.

9. Set realistic stretch goals for each metric. Make them tougher than 'business as usual' but avoid 'pipedreams'.

10. After deployment, remain vigilant for data trends that might mean data isn't being collected properly or a metric has become obsolete.

11. Display metrics data prominently and emphasize it at every meeting so that everyone in the organization is constantly aware of its importance and implications.

Figure 6.3   Summary of metrics development tips

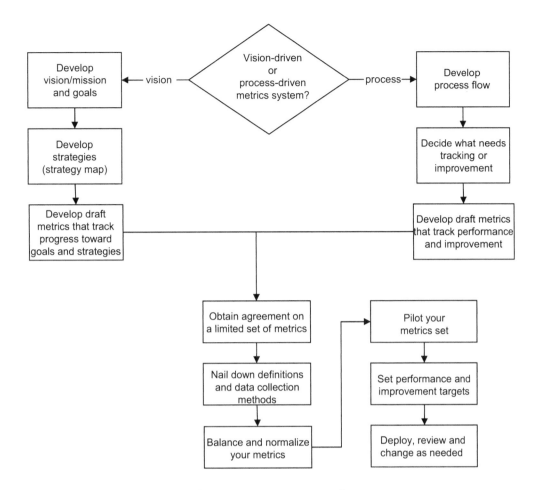

Figure 6.4 A decision tree for developing your metrics system

CHAPTER 7

Displaying Your Metrics

I have a friend who owns a manufacturing and assembly company, providing parts for the aerospace and automotive industries. It's a mid-sized company, with about $100 m in sales and several manufacturing locations around the US. Some years ago they won the Malcolm Baldrige National Quality Award – the highest quality award given in the US. The company did a lot of things right to win the award, but one of the things that struck me most was the way they displayed data. It was simple and ingenious.

Each of their production lines has three to four quality and delivery metrics that are tracked on a daily basis. All of these metrics are posted on the walls of the cafeteria, using a run chart (performance vs. time) for each metric. Next to each chart is a flag holder that contains either a red or green flag:

- If the metric value is OK at the moment, it has a green flag.
- If there are problems (product quality is below expectations or delivery dates are slipping), it has a red flag.

Each shift, the entire plant assembles in the cafeteria for a quick status review. Any metrics that have red flags get immediate attention. If the flags are green, it's assumed that the team performing that work has everything under control. The goal of the meeting (never more than a few minutes long) is to work together to ensure that all of the flags are green. Because performance is tracked daily, it's hard for any metric to get too far out of whack before it's dealt with.

I used a similar method on a complex development program in the early 1990s. In that system, a map of the entire development process was posted on a conference room wall. One or two metrics were listed for each process step and each metric was coded with a large blue, green, yellow or red dot:

- Blue meant things were ahead of target for that step.
- Green meant the metric was on target to achieve the goal.
- Yellow meant that the metric was either slightly behind target or was trending in the wrong direction. Even if performance was on target but the trend over time was negative, that metric was coded yellow.

- Red meant things were not going well at all. Performance was behind target and the trend was probably negative as well. These metrics needed remedial attention fast.

The project manager held daily 'stand-up' meetings in the conference room – everyone remained standing so that no one got too comfortable and the meetings never lasted more than 15 minutes. The entire discussion was driven by the coded metrics. Metrics that were blue or green were generally ignored, and attention was focused on changing the yellow and red metrics to green or blue. Particular attention was paid to the yellow metrics so that we could make sure they never went red.

In both of these cases, the organization's work was managed on a day-to-day basis using a limited set of key metrics. The display techniques focused everyone's attention on those matters that needed fixing. Simplicity and constant visibility were key components in both approaches.

THE IDEAL DISPLAY SYSTEM

The ideal metrics display system has the attributes shown in Figure 7.1. First, it provides appropriate visibility for all levels in the organization. On the one hand, it must provide high-level, at-a-glance visibility for executives. On the other, it must provide detailed performance tracking for managers and line personnel. Providing both levels at the same time would seem to be impossible. However, both of our example systems did just that. The colored flags and dots provided anyone surveying the wall with an instantaneous overview of performance for the entire organization. Someone who had never before seen the system could instantly judge whether things were going well or poorly simply by observing whether the colors on the wall were predominantly green or red. Meanwhile, managers and line personnel were immediately drawn to those flags and dots that were yellow or red. There, trend data and explanations provided detailed information about the current state of these processes and products, along with explanations and corrective actions.

Visibility

      Provides high-level, at-a-glance visibility for executives
      Provides detailed performance tracking for managers
      Located where all employees can see it
      Is reviewed daily by all affected employees

Information

      Contains a snapshot of up-to-the-minute information
      Signals potential problems
      Provides trends
      Indicates corrective actions planned or underway

Figure 7.1   Attributes of an ideal metrics display system

Both of our example systems were displayed prominently on the wall in a highly visible location. In each case, management chose to conduct daily coordination meetings directly in front of the metrics display, which forced everyone to view the metrics every day.

In contrast to the ideal system, most display systems exhibit a number of common problems (see Figure 7.2). Most display systems aren't easily accessible or constantly in view. They are hidden in a report or a computer system. The most egregious examples are those systems that generate large reports that are routed to only a select group of people. Many companies collect the data from across the company, but distribute the results on a 'need to know' basis. The metrics systems themselves are maintained by – and accessible to – only a small group of people whose job it is to maintain the system. Executives restrict access to key performance or financial data on the assumption that this is executive-level information that is not appropriate for lower-level personnel. Yet day-to-day decisions at all levels of the organization are driven by these numbers! In effect, by restricting visibility of the information but making key operational decisions based on it, management is constantly surprising the organization. Lower-level managers and line personnel who are not privy to this need-to-know information never see the data on which the decisions are based, so those decisions appear random and arbitrary. This is the old Taylor management model that assumes that line workers aren't smart enough to make any decisions and must be directed by managers at every turn. Modern organizational theory has rejected the Taylor management model in favor of participatory, team-based systems that push decision making to the lowest possible level. However, Taylor lives on in many of these 'need to know' metrics systems.

Location
- Isn't easily accessible to all employees
- Isn't constantly in view
- Isn't referenced in daily or weekly meetings

Understandability
- Hard to understand at a glance
- Visuals not compelling

Communication
- Doesn't highlight key issues
- Doesn't have current, up-to-the minute information
- Doesn't show trends
- Doesn't motivate

Figure 7.2   Problems to avoid in display systems

One might think that putting the metrics in a computer system that is accessible to all employees is ideal; anyone can access the data whenever they want and it's easy to keep the information up-to-date. However, this approach actually results in an 'out of sight, out of mind' mentality. I'm sure you've experienced the situation where electronic information is constantly sent to you (through websites or emails), but you rarely look at it unless you have a pressing need because you just have so much other stuff to deal with. The same is true for metrics system data. Yes, it's easy to access from the computer, but no one ever logs on unless they have a specific need, such as to enter data. Unless the metrics data is in a place where it's regularly seen and evaluated, most people will simply ignore it. Computer systems are wonderful collection, analysis and storage systems, but they're very poor display systems.

An additional problem with computer or report-based display systems is that it's too easy to add to or modify the metrics. The result is that these display systems tend to get more complex, voluminous and sophisticated over time and they end up obscuring the key messages. With a simple wall-chart system, any modification is labor-intensive, so those maintaining the system will resist changes that aren't critical to the display.

Putting a display up on the wall also invites people to critique it. If the information isn't easily understandable, your staff will tell you about it. If it's not compelling, they'll complain or perhaps ignore it. On the other hand, if it's clear, compelling, and it highlights the issues, you'll find people clustering around the display and discussing it.

The fundamental message here is: keep it visible, simple, compelling and up-to-date.

So, what do you do if you've got a multi-location organization with facilities around town, around the country or around the world? I worked with one client where we had literally dozens of locations where the metrics had to be posted. We created a metrics board that every location could mount on a prominent wall. The board had 12 Plexiglas slots, one for each of the key metrics. The database administrator generated the weekly metrics plots and then emailed them to each location, where one person was designated to print out the metrics sheets and put them in the Plexiglas slots. One executive had responsibility for periodically checking on the different locations to make sure they were updating the metrics sheets and using the data. The net result was that each location had full and constant access to the same metrics data. A different client did essentially the same thing, but used their existing closed-circuit TV system to display metrics data in every cafeteria and building entryway.

TYPES OF DISPLAY SYSTEMS

Let's look at several different types of display systems.

Basic run chart displays

The simplest display system is a set of run charts showing performance over time. Figure 7.3 shows a basic run chart, with performance data and a goal. In this case, performance is not meeting the goal, but it is slowly improving – with the exceptions of February and March. This metric would probably merit a yellow flag (although in February it would have been red).

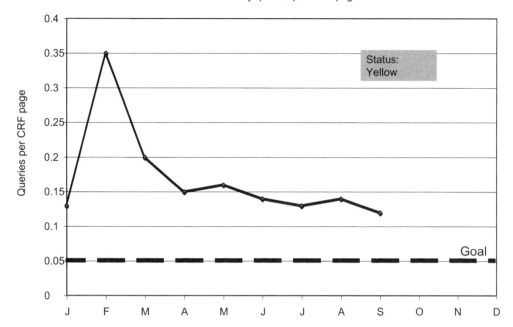

Monthly queries per CRF page

Figure 7.3 A simple run chart

More sophisticated run chart displays

A slightly more sophisticated display system augments the run chart with accompanying key information, as shown in Figure 7.4. This display is more descriptive and allows casual observers to understand both what is happening and what is being done to make improvements. While the simple run chart in Figure 7.3 is sufficient for teams that constantly view and work with the metrics data, the more sophisticated chart in Figure 7.4 works better in situations where the data is referenced less frequently – say when the team only gets together once a week or once a month.

A full set of these run charts would typically be displayed on a single wall with flags or dots to provide a complete view of the project or organization status.

Other types of charts – bar charts or box-and-whisker charts – may be used instead of the line charts shown in Figures 7.3 and 7.4. Figure 7.5 shows the same data displayed as a box-and-whisker chart. In this case, the query data for each trial is used as the basis for the plot. The box represents the range from 25th to 75th percentile for all trials operating during the month, while the whiskers show the best and worst trials during the month. This representation shows that at least one trial has been achieving the goal of 0.05 queries/CRF page every month since March. This information – which wasn't available in the simpler run charts – might lead the organization to look for best practices.

In these types of systems, it's best to have one person who is responsible for maintaining the system (a database administrator) and then an owner for each metric. If the data collection system is intranet-based, then each metric owner can input the data and print out the performance charts. The database administrator is responsible for two things:

Status: Yellow

Owner: John Doe

Metric definition:
Total queries generated during the month for all clinical trials divided by total CRF pages entered by data management for all trials during the same month.

Included/excluded data:
Phase II, III and IV queries included.
Phase I queries excluded.
Type I queries included. Type II queries excluded.
Paper CRF queries only.

Data sources:
All data from Data Management.

Data integrity issues:
None

Reporting frequency:
Monthly

Goal:
0.05 queries per CRF page on a monthly basis by December.

Comparison:
Best paper CRF query rate found in the literature is 0.06

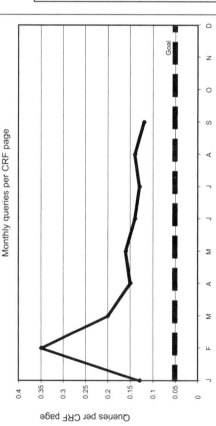

Monthly queries per CRF page

Trends: Improving

Probability of achieving goal: Low

Improvement strategies:
✓ Instituted new site training method last December.
✓ Began measuring CRAs on their site query rates in March.
✓ Began sending query rate data to sites in August.
  Will begin following up with sites on query reports next month.

Variances:
February and March data were worse due to launch of a trial in a new therapeutic area where we were using new sites.

Corrective actions for variances:
None

Figure 7.4 A more sophisticated run chart

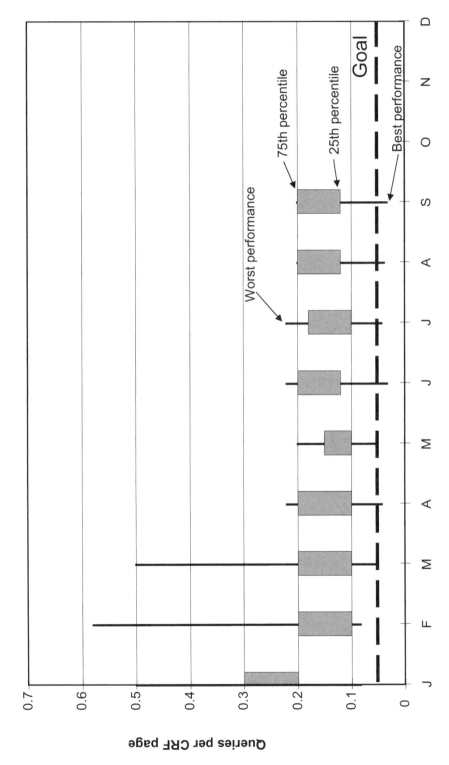

Figure 7.5 A box-and-whisker plot

1   Ensuring that the collection system is functioning properly, and
2   Helping metric owners to correct any problems they are having entering data and
    printing out their display charts.

Because these are fairly simple charts, a large cadre of people to maintain the metrics
system isn't required. Managers will certainly complain that they don't have enough time
to maintain and print out their metrics, but this is generally a specious complaint. Basic
charts such as these can all be preprogrammed into general-purpose software such as
Microsoft Access or Microsoft Excel with minimal effort. They can also be easily derived
from off-the-shelf metrics system software such as Performancesoft's pbviews, Pilot
Software's PilotWorks, Oracle's SEM, or Cognos' Metric Manager (and there are many
other providers as well).

All of our example companies could use this type of display system.

Process map displays

A slightly more complex system ties the metrics to a process map and keys the data display
to the map. Figure 7.6 shows a generic flow chart where the various process steps are
color coded according to their current status. Green boxes are complete; red boxes are on
the critical path and behind; yellow boxes are behind but not on the critical path; and
white boxes haven't started yet. This display gives everyone an immediate view of the
status of the entire flow. Attached to this flow are a series of metrics that monitor the status
of key parts of the process, such as protocol development or CRO performance. These
could be the same run charts discussed previously, or they could be only those metrics that
require attention.

In this system, the emphasis is on the process, as well as on recovery actions necessary to
keep the process on schedule. *Inventrix* could use this method to track the performance of
their innovation process. *BioStart* could use this to track the performance of each clinical
trial. In *BioStart's* case, this system would be particularly effective in keeping management
apprised of what's going on in each trial at all times.

Spider diagrams

A spider diagram is the most sophisticated display system we'll discuss. It combines data
from many different metrics into a single diagram. It provides observers with an overview
of what's going on, but doesn't provide a great deal of detail. Figure 7.7 shows a spider
diagram for an organization like *VirtuPharm*, which is trying to improve its relationships
with CROs. In this particular diagram, the pharma has settled on 16 different metrics. The
metrics have been divided into four categories:

1   Project performance (metrics 1–4)
2   Teams (metrics 5–8)
3   *VirtuPharm*–CRO relationship (metrics 9–12)
4   Site performance (metrics 13–16).

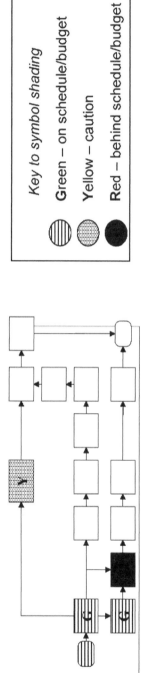

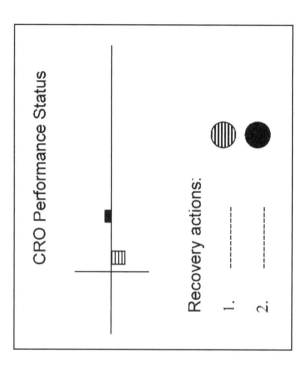

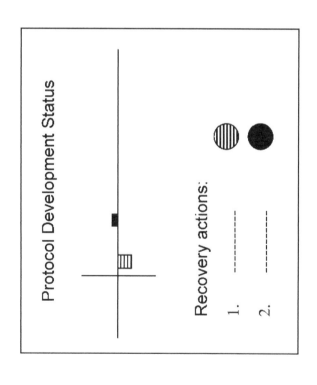

Figure 7.6 Schematic for a process map performance system display

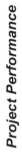

Project Performance

1. Project setup cycle time
2. Project completion cycle time
3. Enrollment efficiency
4. Project cost

Teams

5. Team performance index
6. Pharma team turnover
7. CRO team turnover
8. Team meeting quality

VirtuPharm–CRO Relationship

9. % of projects with preferred CROs
10. % of repeat business
11. Pharma team experience level
12. CRO team experience level

Site Performance

13. EDC usage
14. Cutting edge systems usage
15. Average site effectiveness
16. % of sites achieving 90% of enrollment target

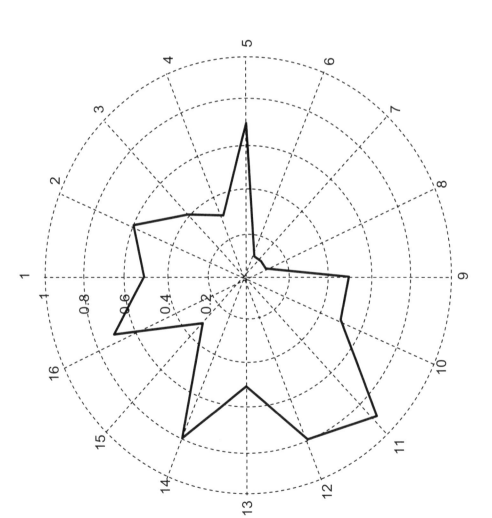

Figure 7.7 Spider diagram display system for a pharma–CRO relationship

Each of the 16 metrics measures performance across all of the projects and all of the CROs being used by the pharma. Progress is reported on a quarterly basis, although some metrics – such as metrics 1–4, 15 and 16 – can only be measured once for each clinical trial. For instance, let's assume there are 100 ongoing clinical trials. Twenty-five of them completed this last quarter, and 30 new ones started up.

- The 25 completed trials could be measured for the four project performance metrics (setup cycle time, completion cycle time, enrollment efficiency and project cost) plus metrics 15 and 16.
- All 100 projects would be measured for the other 10 metrics.

So far, so good. However, spider diagrams present a bit of a challenge at this point. As you can see from Figure 7.7, each metric is plotted on a radial line from the center of the plot out to the outer circle. The distance along each radius is the same, and is marked off from 0.0 (at the center) to 1.0 (at the outside). For our spider, we'll assume that 0.0 is the worst performance and 1.0 is the best performance.

Unfortunately, our 16 metrics don't have ranges from 0 to 1:

- The cycle time metrics can run from negative to positive. If they're measured in days, the range can be perhaps from –60 days to +200 days, or more.
- Meanwhile, the project cost is measured in dollars and can run into the millions.
- The percentage metrics run from 0 to 100.
- Index metrics like team performance index (metric 5) or average site effectiveness (metric 15) typically use a 5- or 10-point Lickert scale.

In order to fit all of these metrics on the same diagram, we have to do something to force the range of each metric to be between 0 and 1. We do this first by expressing everything we can as a fraction:

- We can easily express percentage metrics as fractions.

- We can compress a 5- or 10-point Lickert scale to a 0–1 scale by dividing by the size of the scale (5 or 10).

- Other metrics can be expressed as ratios of actual value to contracted or forecast value. Metric 4 (project cost) can be expressed as a ratio of actual cost to contracted cost. We saw in Chapter 6 that this was a good idea anyway, since expressing the cost this way allows us to compare large projects with small projects. We can also express cycle times as a ratio of actual value to contracted or forecast value.

This helps quite a bit, but it's not sufficient. While many metrics are now on a 0–1 scale, the cost and cycle time metrics could still have ranges from 0 to perhaps 10 or more. If we measure the ratio of actual cycle time to forecast, the actual could be quite a bit larger than the forecast, creating a metric value which is greater than 1.0. For these metrics, we use something called a utility function. Simply put, the utility function allows us to plot a given metric of indeterminate range on a 0–1 scale. Figure 7.8 shows how we can create a

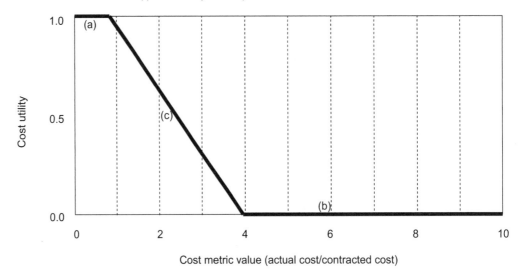

Section (a) has a cost utility of 1.0 for any actual cost that is less than the contracted cost.
Section (b) has a cost utility of 0.0 for any actual cost that is more than four times the contracted cost
Section (c) has a cost utility that linearly decreases from 1.0 to 0.0 as the cost metric rises from 1.0 to 4.0

Figure 7.8 A utility function for project cost

utility function for a cost metric. First we put down what we believe is the acceptable range for the metric, which in this case is the ratio of actual to forecast cost:

- We know that any metric value less than 1.0 (actual cost is less than contracted cost) is good.
- And let's say that we as an organization believe that any metric value greater than 4.0 (actual cost is four times greater than contracted cost) is totally unacceptable.

Then we can create a utility function that translates the cost metric to the cost utility:

(a) translate any cost metric value less than 1.0 to a cost utility value of 1.0 (maximum value)
(b) translate any cost metric value greater than 4.0 to a cost utility value of 0.0 (no value at all)
(c) translate any cost metric value between 1.0 and 4.0 linearly to a cost utility value between 1.0 and 0.0 (steadily decreasing value). We could also have used a nonlinear relationship, but we'll keep it simple for our example.

Utility functions will allow you to map any metric onto a 0–1 scale. You can also reverse the function, making 0.0 equal desirable performance and 1.0 equal poor performance.

Once you've normalized all of your metrics to between 0 and 1, they can all be plotted on a spider diagram. Just make sure that all of your metrics have better performance at the same end of the scale; either best performance is at the origin (0.0) or best performance is at the outer circumference (1.0). Otherwise, everyone looking at your diagram will get confused about which direction is better and which is worse.

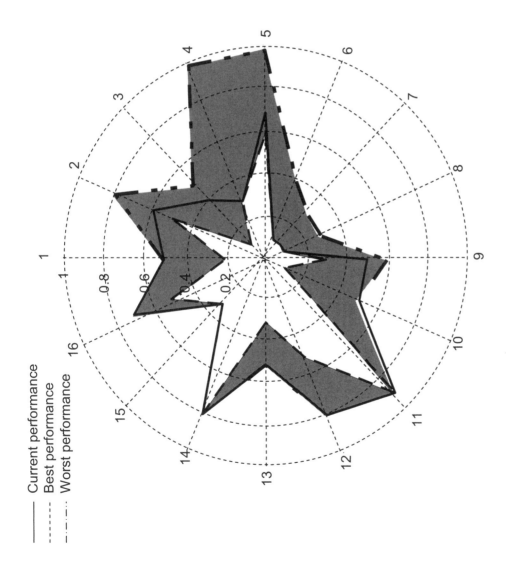

Project Performance

1. Project setup cycle time
2. Project completion cycle time
3. Enrollment efficiency
4. Project cost

Teams

5. Team performance index
6. Pharma team turnover
7. CRO team turnover
8. Team meeting quality

VirtuPharm–CRO Relationship

9. % of projects with preferred CROs
10. % of repeat business
11. Pharma team experience level
12. CRO team experience level

Site Performance

13. EDC usage
14. Cutting edge systems usage
15. Average site effectiveness
16. % of sites achieving 90% of enrollment target

— Current performance
----- Best performance
—-—- Worst performance

Figure 7.9 Spider diagram showing best and worst performance

You can make your spider diagram more sophisticated by adding best and worst performance to the graph. For instance, Figure 7.9 shows the same set of 16 metrics, with current as well as best and worst performance over the last eight quarters shown for each. In this chart, closer in to the center represents better performance, while farther out represents worse. The shaded area shows the range of performance over the last eight quarters. Thus:

- Metrics 1, 10, 12, 13 and 16 are at their recent nadirs (larger values indicate worse performance).
- Metrics 4, 6, 7 and 8 are at their zeniths (smaller values indicate better performance).
- Metrics 11, 14 and 15 have seen no change in performance over the last two years (either for better or worse).
- The remaining four metrics are somewhere in between.

SUMMARY

In this chapter, we've discussed how to display metrics in a way that makes people pay attention to them and act on them. Simple wall charts are the best approach. If you've got a multi-location organization, set aside a wall at each location to put up the charts and distribute them via email or the web. Figures 7.1 and 7.2 provide attributes of good and bad metrics systems.

We also discussed various types of display systems, including run charts, box-and-whisker plots, process map displays and spider diagrams. You'll have to use trial and error to figure out which display type is best suited to your metrics and your organization. In all cases, the goal is to display your data in a simple-to-read manner.

CHAPTER 8

Closing the Strategy Loop

Since the mid-1990s, I have been promoting the value of metrics to the pharmaceutical R & D community. From my vantage point as a process improvement consultant, it seemed obvious that it was impossible to improve something you couldn't measure. If we don't know how well we are doing or what our goals are, there's no way to make significant, meaningful improvements. By 2000, I had persuaded the Institute for International Research of the importance of metrics in R & D and they offered to hold an annual R & D Metrics conference. I have been fortunate to chair that conference in the US for the last four years. The conference has now spread to Europe and I had the opportunity to chair the European conference in 2005. It was at the most recent of these conferences that the importance of tying R & D metrics to R & D strategy became clear yet again. Gen Li of Pfizer was reviewing the progress in R & D metrics over the past decade or so.[1] He noted that a key metric for every large pharma R & D organization has been the number of new chemical entities brought to market every year. This has often been expressed as 'bring two to three blockbuster drugs to market each year'. Such a goal seemed financially doable, since a typical drug costs less than $1 b to bring to market and the annual industry R & D spend has been growing rapidly (it totaled more than $200 b over the last ten years[2]). Past history had shown that bringing one or more blockbusters to market each year was reasonable. Two or three per year seemed a huge stretch, but not impossible. But let's look at the implications of this goal:

- If each of the top ten pharmas successfully execute this strategy, 20–30 blockbusters will come to market each year.

- Only 3 in ten newly marketed drugs ever have sales that exceed even the average R & D costs to bring them to market,[3] and only a fraction of the drugs that do come to market are blockbusters. So 20–30 blockbusters means that the top ten pharmas would have to bring at least 67–100 new drugs to market each year.

1. G. Li, 'Bridging Strategic Metrics and Operations Metrics', IIR Pharmaceutical Performance and Metrics Conference, January 24–26, 2005.
2. 'Pharmaceutical Industry Profile 2004', www.phrma.org, p. 7.
3. H. Grabowski, J. Vernon and J. DiMasi, 'Returns on Research and Development for 1990s New Drug Introductions', *Pharmacoeconomics* 20, suppl. 3 (2002): 11–29.

- 1000 compounds are typically discovered for each drug that makes it to market. Therefore, as many as 100 000 new compounds need to be discovered per year to support this goal, just by these ten pharmas!

In fact, instead of 100 new drugs to market by just the top ten pharmas, an industry-wide average of only 32 new medicines were approved by the FDA per year over the last decade![4]

Gen Li's point was this: while every company professed a goal of getting these blockbusters to market, not a single company had succeeded even once, mainly because this metric had never been 'operationalized'. The operational strategies and metrics required to make such a goal reality were never put in place. Instead, organizations simply looked at the goal and said 'let's simply do a *lot more* of what we're doing!' and hoped that blockbusters would come out the end of the pipeline.

The systematic linkage of goal to strategy, strategy to action, and action to deployment, with metrics as the feedback mechanism was never implemented (see Figure 8.1). The industry never linked the goal of 'two to three blockbuster drugs a year' to operational strategies that could make the goal achievable and which could then be systematically deployed and measured. (Note: Wyeth is now pursuing this goal with a systematic strategy they call '12 in 2 out'.[5])

Throughout this book, I've emphasized the importance of setting goals, building strategies that link to the goals and then designing metrics that will measure the success of those

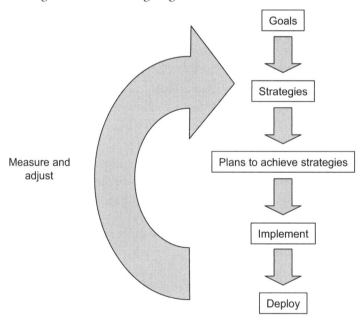

Figure 8.1 Successful strategy implementation requires linkage between goals, strategies, action, deployment and metrics

4. 'Pharmaceutical Industry Profile 2004', www.phrma.org, p. 11.
5. K. Ghosh, 'The Metrics Behind Wyeth's Increased R & D Productivity', ExL Pharma Pharmaceutical Performance Metrics and Benchmarking Summit, October 19–20, 2005.

strategies. I'd like to spend this last chapter discussing how to use these techniques to drive real improvement in your R & D organization. Building strategies and metrics by itself isn't enough. You have to successfully implement and deploy those metrics and strategies to be successful.

DEPLOYING METRICS THROUGHOUT THE ORGANIZATION

The key to deploying metrics throughout your organization is the line-of-sight concept we alluded to in Chapter 3 (see Figure 8.2). In order to keep everyone in the organization aligned and focused on the goal, we need to make sure that every group and person in the organization can clearly link his or her behavior and performance to that goal. The problem is that individual employees down in the trenches often have trouble relating to ('seeing') a top-level organizational goal. Here's an example: A bench chemist doing day-to-day HPLC analyses has trouble relating this work to an organizational goal of increasing profit or decreasing time to market. She is simply doing what she was hired to do. If department procedures are cumbersome or customers ask for unnecessary analyses that simply confirm what is already well known, she may feel she has little power or incentive to improve things. However, by creating local, line-of-sight goals, we can localize the organizational goals. For instance, we can give both the local department and the individual chemists a local goal of ensuring that every analysis is done accurately and to customer specifications each and every time. We could go even further, and give that

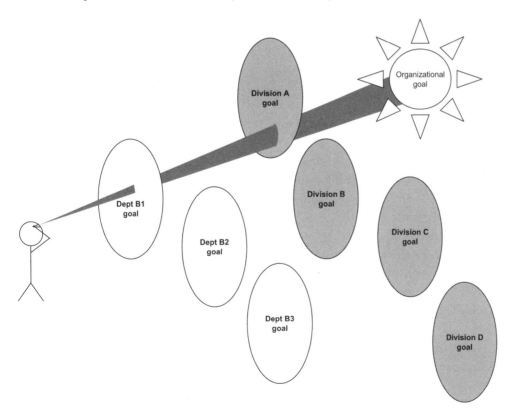

Figure 8.2   Line of sight from individual to goal

chemist some ability to question whether a given analysis should be done at all (for example, is it just confirming what we already know, as opposed to yielding significant information). Thus, we might end up with four key measures for that chemist's lab:

1   Percentage of analyses that were simply confirming other findings
2   Percentage of analyses that meet customer expectations and quality goals the first time
3   Average cost per analysis
4   Average cycle time per analysis.

We now have a balanced set of metrics at the local level that link to such high-level

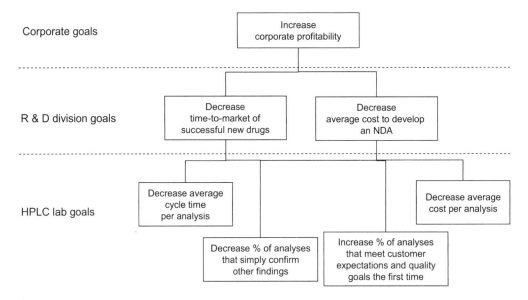

Figure 8.3   How organizational profitability goals can be linked to individual and local department operational goals

metrics as profitability and time to market. These linkages are shown in Figure 8.3.

In general, then, we need to take the set of top-level, organization-wide metrics and cascade them down through the organization, creating local metrics at each level that clearly link to the metrics at the level above (see Figure 8.4). This can be done in one of two ways (see Figure 8.5):

1   We can cascade the strategy maps down through the organization by having each lower-level division create its own map. In this approach, each lower-level division takes one or two strategies from the higher-level map and adopts them as its goals. It then creates a new strategy map and develops metrics that tie to the map.

2   We simply cascade the metrics themselves down to successively lower levels of the organization, and ask each successive level to develop local metrics that will tie directly to one or more of the metrics at the level above.

Using this cascading approach to metrics has an additional advantage. We've already seen that it creates alignment vertically in the organization through line-of-sight metrics. But it

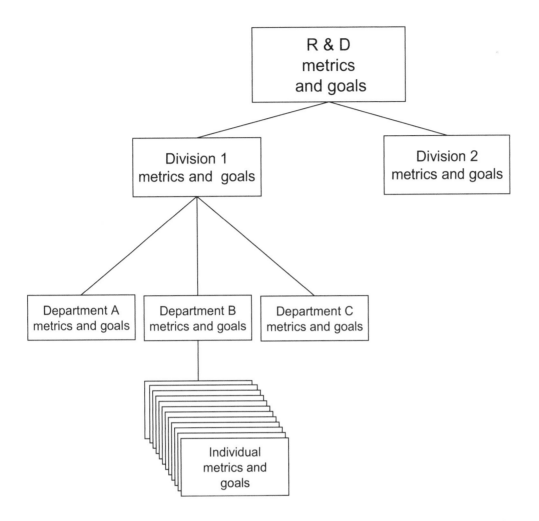

Figure 8.4 Top-level goals and metrics must cascade to successively lower levels in the organization

also creates alignment horizontally across departments and divisions. It turns out that horizontal alignment is critical to organizational improvement, and lack of it can cripple performance. Here's why: In attempting to meet organizational targets for increased profitability or cost reduction, departments will often attempt to optimize their own performance at the expense of the entire organization. For instance, a shared resource department such as an HPLC lab may try to reduce staff in order to increase the productivity of each chemist. But this means longer cycle times and/or lower quality, since there are fewer people to handle the workload. The lab may have saved money in staffing costs, but at the same time may have cost the organization much more in longer times to market or decreased efficiencies in other departments that have to wait for the results. When strategy maps and metrics are cascaded down through the organization, senior management has the ability to look at the cascades and identify potential horizontal conflicts before they adversely impact performance.

In general, cascading strategy maps (Figure 8.5a) is a more robust approach than cascading metrics (Figure 8.5b) because the linked maps provide more visibility into:

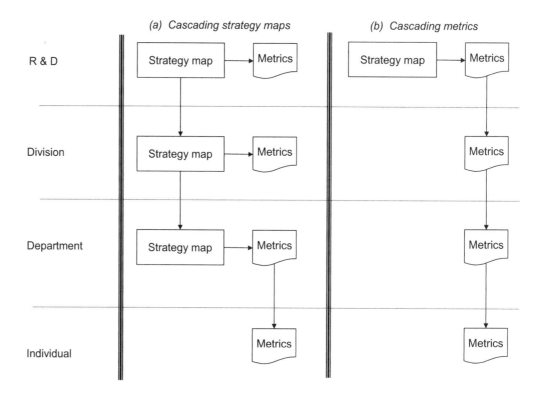

Figure 8.5 Two possible methods for cascading metrics down through the organization

- which top-level strategies are not being adequately addressed by the organization
- which top-level strategies are being overemphasized
- which local strategies may be overlapping or conflicting with each other (for example, across departments).

However, it is also more complex and time-consuming.

USING METRICS TO IMPROVE PERFORMANCE

Once your metrics are fully deployed and cascaded, you can begin to use those metrics to improve organizational performance. Figure 8.6 shows a typical approach for making improvements in processes and organizations. It involves five steps.

First, start by creating your metrics and measuring the performance of key parameters. We've spent the entire book discussing this one, so no further discussion is required here.

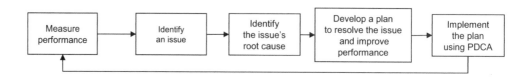

Figure 8.6 Typical improvement process

Second, identify those metrics that are lagging behind your goals or are perhaps deteriorating. If you've developed targets and tracking systems as we discussed in Chapter 7, it will be easy to identify the metrics where improvement is required. You should also be able to get agreement from your entire organization as to which metrics are most important to address, since everyone is looking at the same display system and observing the same trends. It's critical to get this agreement at the very beginning of your improvement process so that you can avoid conflicts and minimize resistance later on. In every improvement project, there's a point where one group or another in the organization begins to get very nervous. Whether it's real or perceived, these people feel threatened by the increased attention or the pending change, and they will work to slow down or stop the process. They may even try to withdraw from the improvement project altogether. If there is strong, organization-wide support for the project up front, you'll be able to deal effectively with the recalcitrant group. However, if up-front agreement is weak, you may not have enough organizational support to overcome this resistance. I've found that you have the best chance of getting strong, organization-wide support for your improvement projects if you follow these steps:

1   Assemble your senior management team.

2   Present and discuss the current status of your metrics and the merits of making improvements to each.

3   Use a multi-voting technique to allow each management team member to vote for the two or three metrics they believe will create the most value if improved.

4   Decide on how many projects to undertake (definitely not more than three) and select the metrics that receive the most votes.

5   Instruct the management team immediately to write a brief improvement charter for the selected improvement projects, including goal, anticipated benefits from improvement, improvement time frame, senior management sponsor, improvement team leader and membership.

This approach maximizes the amount of support you get from your organization and minimizes the risks of resistance during the improvement process.

Third, having identified the metrics requiring improvement, you now have to figure out what's causing the problem. You can use any number of process improvement approaches to do this. The most effective approaches include these steps:

1   A detailed documenting of the existing process, including the inputs to and outputs from each task.

2   Identification of all the customers and suppliers for the process and for each process task.

3   A review of your process map by your team, as well as by customers and suppliers, to

determine which parts of the process work well and which cause problems. An easy way to do this is to compare the outputs of each task to the required inputs for the next task and look for misalignments or 'disconnects' (an approach called 'next operation as customer').

4   A failure modes and effects analysis to determine the frequency and consequences of each process disconnect. From this analysis, you can rank order the disconnects and failure modes and select the critical few.

5   A root cause analysis of the critical few disconnects and failure modes to determine the root cause of each. I find that pharmaceutical R & D organizations respond best to the '5 why's' root cause analysis technique.

Fourth, you can now develop improvement plans to increase the performance of your process and improve your metrics. You'll need to use some creative thinking approaches to encourage your team to think about novel solutions, because most of your team members are likely to be highly skilled at executing the process in the traditional manner. Team members may acknowledge that the process is not working as well as it could, but it will be very difficult for them to think of completely new ways of doing things. On the other hand, pharmaceutical R & D staff are typically highly educated and fundamentally inquisitive. Once they've learned several creative thinking techniques, I find that they are quick to apply these techniques to the process and will come up with a number of clever ways to improve things.

Once your team has developed an improvement plan, you need to go back to the senior management team and obtain approval for the change. This is a critical step, because you'll need the management team's support when you start the implementation.

Fifth, you can start the implementation. A good implementation approach should use a Plan-Do-Check-Act (PDCA) strategy:

Plan a specific implementation step
Do the implementation in a small area of the organization (a pilot)
Check the implementation to make sure it was successful. If not, make appropriate changes and redo the pilot
Act on the successful pilot and spread the change across the organization.

Make sure you work carefully to spread the new approach with a minimum of resistance. Use a variety of communication methods and peer-to-peer contacts to minimize the amount of resistance to the change.

SUMMARY

In this final chapter, I have attempted to close the strategy loop by linking top-level metrics both down through the organization and back up to strategies through improvement actions. To do this you must first cascade your metrics down through the organization using either cascading strategy maps or cascading metrics (see Figure 8.5). Remember to create lines of sight from individuals and departments all the way up to your top-level R & D metrics. Then make sure that all of the various department metrics are aligned and none of the departments are working at cross-purposes.

When you've done this, use the approach shown in Figure 8.6 to make improvements that will move your metrics toward their goals. Use tried-and-true process improvement methods and the Plan-Do-Check-Act cycle to do this successfully.

Once you've accomplished these actions, you'll be well on your way toward making metrics a driving force for organizational improvement.

Index

tracking systems
  clinical trials 12
  metrics systems 20, 24–5
  use 24–5

utility functions, project cost 116

VAC 78

VirtuPharm 12-13, 28, 44–6, 64, 66–7, 90–91
vision driven metrics systems 96

weakest link, development 33
work
  effective/ineffective 2–5
  smart 2

About the Author

David S. Zuckerman is the world's foremost expert in the area of pharmaceutical R & D metrics, and lectures internationally on pharmaceutical metrics and balanced scorecards. He is owner and principal of Customized Improvement Strategies LLC – a management consulting firm that focuses on process improvement, organizational development, leadership development and change management; his clients include large and small pharma R & D organizations, CROs and academic medical centers. Find out more about his work at www.rx-business.com.